WILDERNESS PRESERVATION

A Reference Handbook

WILDERNESS PRESERVATION

A Reference Handbook

Kenneth A. Rosenberg

CONTEMPORARY WORLD ISSUES

ABC-CLIO
Santa Barbara, California
Denver, Colorado
Oxford, England

Library of Congress Cataloging-in-Publication Data

Rosenberg, Kenneth A.
    Wilderness preservation : a reference handbook / Kenneth A.
Rosenberg.
        p. cm. — (Contemporary world issues)
    Includes bibliographical references and index.
        1. Nature conservation—United States—Handbooks, manuals, etc.
    2. Wilderness areas—United States—Handbooks, manuals, etc.
    I. Title. II. Series.
    QH76.R66   1994   333.78′216′0973—dc20   94-19336

ISBN 0-87436-731-X (alk. paper)

00 99 98 97 96   10 9 8 7 6 5 4 3 2

ABC-CLIO, Inc.
130 Cremona Drive, P.O. Box 1911
Santa Barbara, California 93116-1911

This book is printed on acid-free paper ∞ .
Manufactured in the United States of America

To my parents,
Richard and Karen Rosenberg

Contents

List of Figures and Tables

Figures

Tables

Preface

WHEN EUROPEAN EXPLORERS FIRST came to North America in the sixteenth century they found a continent of vast resources virtually untouched. These men were struck by the enormous potential of the land for farming, plentiful wild game, and a seemingly endless supply of wood from forests stretching into oblivion. The land was so immense that the first major overland expedition to the far West did not even begin until Lewis and Clark set out in 1804, almost 300 years later.

To early settlers the thought of depleting the land of trees would have seemed preposterous, but by the end of the nineteenth century concerns were beginning to mount. Now, 500 years after Columbus, most of the original forests have fallen victim to logging, development, roads, and dams. Wetlands, free flowing rivers, and animal habitats across the country are also slowly diminishing.

Opinions on how to solve today's wilderness controversies vary widely. Most agree that at least some wilderness management is necessary, but how much protection to give our open lands is in constant debate between preservation-minded environmentalists and use-oriented business leaders.

Over the years, the ideology behind managing resources in the United States has evolved to reflect the needs of the nation. Agencies such as the Department of the Interior and the U.S. Forest Service were formed by a nation trying to best divide its resources to create a powerful industrial economy. Today these agencies support the demands of the economy, but also manage the resources that remain in order to strike a balance between use, conservation, and preservation.

This book will present the issues and the history of wilderness preservation and conservation in the United States, and provide sources for further study of the topic. Chapter 1 is an introduction to wilderness in the United States and how it is managed, with a look at the major issues and controversies surrounding our remaining open spaces. Chapter 2 is a chronology of events that have had a major impact on our wilderness over the years. Chapter 3 contains biographical sketches of people that have had a large impact on wilderness conservation and preservation. Chapter 4 contains statistical data, quotes, and excerpts from important environmental legislation. Chapter 5 is a semi-annotated listing of protected federal lands, and contains information on the acreage and location of much of the remaining wilderness in the United States. Chapter 6 provides an annotated listing of federal agencies, private organizations, and associations. Chapters 7 and 8 provide print, film, video, and electronic sources of information for further study. Finally, a glossary of terms relating to wilderness preservation finishes the book.

All data in this book relating to wilderness areas and national parks includes figures from the United States Senate version of the California Desert Protection Act. Similar versions of the law were passed by both the U.S. House of Representatives and Senate, but it had not been signed into law as the book went to press. The act creates Joshua Tree, Death Valley, and Mojave national parks, and designates 76 new wilderness areas totaling 7.7 million acres.

Acknowledgments

The author would like to thank the employees in the public affairs departments of the U.S. Forest Service, Fish and Wildlife Service, Bureau of Land Management, and Park Service (particularly Mr. Dario Bard) for all of their help on this project. Thanks also to The Wilderness Society and all of the other environmental organizations, the generosity of the Staude family for the computer, and Barry Dworak for help with the graphs.

1

Introduction

WHEN THE UNITED STATES WAS FIRST established as a country, practically all of North America was one vast wilderness. Early settlers viewed the wilderness as a land that needed to be tamed. Pioneers spread across the country, taking advantage of the vast resources. By the beginning of the twentieth century many of those resources were becoming scarce, and people started to see a need to conserve them. Others saw a need not only to conserve resources, but also to preserve some of the remaining open space for the enjoyment of future generations.

The American wilderness today can be found from the Florida Everglades to Maine's Acadia National Park; from the deserts of the Southwest to Alaska's Arctic National Wildlife Refuge. Wilderness is spread throughout the United States, on both private and public lands, in state and national parks, wildlife refuges, and wilderness areas. Some of the land is owned and protected by the government or by private organizations, while other wilderness has no protection at all.

Conservation and preservation are terms that are confused occasionally. Both are terms that refer to ways of protecting the environment, but not in quite the same way. *Conservation* means to manage an area in order to sustain the natural resources. Replanting trees after an area has been clear-cut is a form of conservation. As the trees grow back, the forest is replaced, though not in its original state. *Preservation* means to protect an area in its pristine, primordial state by setting it aside before mankind has had any impact. Of the two terms, preservation is the most simple: setting

1

FIGURE 1.1

The Public Lands of the Lower 48 States

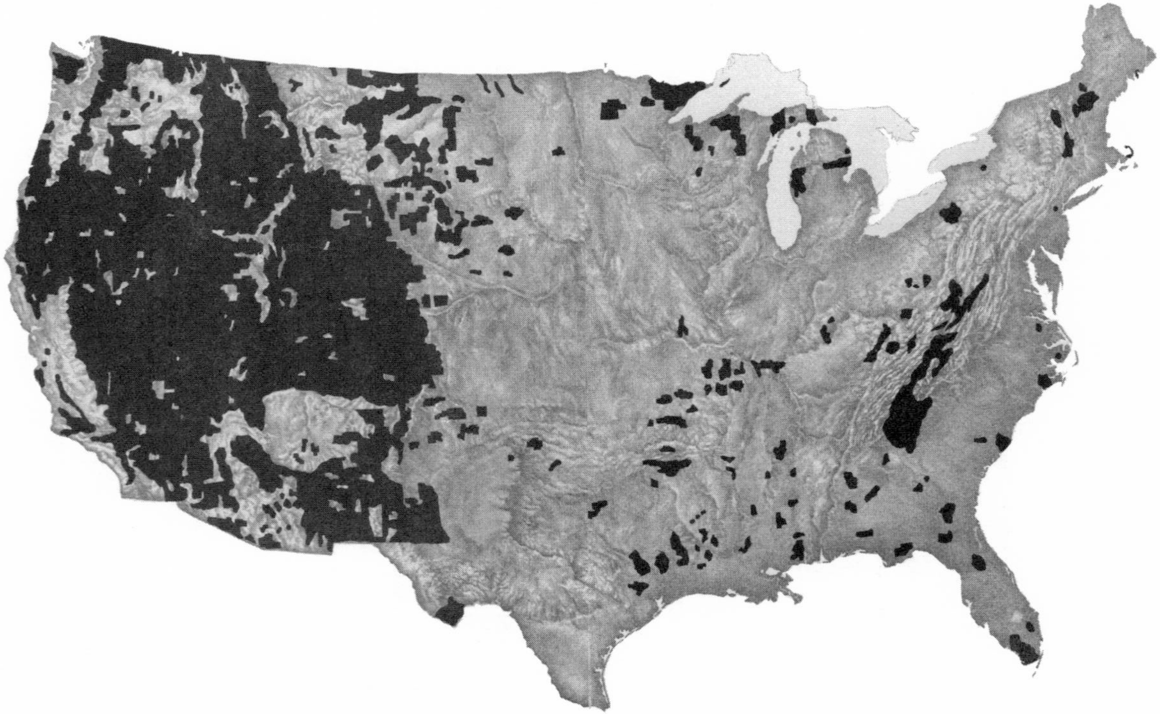

Map courtesy: The Wilderness Society

Note: Map does not include military reservations and miscellaneous other federal lands, as well as 71,000 acres of public lands in Puerto Rico, Guam, the Virgin Islands, and other U.S. territories and possessions.

FIGURE 1.2
The Public Lands of Alaska and Hawaii

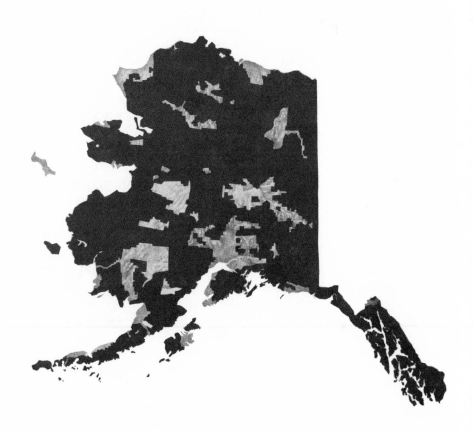

Hawaii

Map courtesy: The Wilderness Society
Note: Maps do not show the Alaska Maritime National Wildlife Refuge—a collection of offshore islands and islets that includes St. Matthew Island in the Bering Sea and most of the Aleutian Islands chain.

aside lands for complete protection. Conservation is broader, and can involve managing the lands in numerous ways. Wolves were intentionally exterminated within the boundaries of Yellowstone National Park. Today plans exist to reintroduce this species to the park. Both the original eradication and the proposed reintroduction are examples of conservation. Preservation would simply have left the wolves where they were in the first place.

Conservationists and preservationists have often disagreed on how to manage specific lands and how much protection those lands should receive. Still other people believe that wilderness should not be protected at all. Many people think that the remaining open lands and resources should not be locked away, but should be put to their most productive use, whether that entails logging, mining, grazing, or farming.

The federal government owns most of the remaining wilderness in the United States, which means that it is up to the American people as a whole to decide how best to manage it. Every square foot of open land is seen as precious by environmentalists, while many who favor development will fight for the right to extract any natural resources that remain in the wild. The wide range of opinions on wilderness issues has led to a multitude of controversies, ranging from logging disputes to disappearing wetlands. This chapter will define the scope of wilderness in the United States, explain how the federal wilderness is administered, and address the major issues and controversies within the scope of land management today.

What Is Wilderness?

The word *wilderness* may have a slightly different meaning to different people, but in general, wilderness is an area unchanged and uninhabited by humans. The U.S. Congress may have said it best when they described wilderness in the Wilderness Act of 1964:

> A wilderness, in contrast with those areas where man and his own works dominate the landscape, is hereby recognized as an area where the earth and its community of life are untrammeled by man, where man himself is a visitor who does not remain.

The size that an area must be for one to consider it wilderness is open to debate, depending upon the area, but the Wilderness

Act specifies that for legal purposes, a wilderness area must be "at least five thousand acres of land or . . . of sufficient size as to make practicable its preservation and use in an unimpaired condition." Several islands designated as wilderness areas are actually less than ten acres.

It is important to note the difference between *wilderness* and *wilderness area*. These two terms are sometimes confused. *Wilderness* is any area unchanged by man, but a *wilderness area* is land that has received designation as such by the U.S. Congress. Congress may designate a specific zone of federal land as a wilderness area in order to give it special protection. Federal public lands with the wilderness area classification become a part of the National Wilderness Preservation System and are completely off-limits to any development or activity which would degrade the area.

National parks, national forests, and wildlife refuges are not necessarily wilderness areas. The designations are completely separate, but often they overlap. National parks are areas administered by the National Park Service (NPS) and set aside to protect their scenic splendor. National forests, administered by the U.S. Forest Service (USFS), are managed to accommodate many uses, including logging, recreation, and preservation. Wildlife refuges are administered by the U.S. Fish and Wildlife Service (USFWS) in an effort to protect wildlife and their habitats. None of these areas are off-limits to commercial development. Wilderness areas *are* off-limits to development, and many national parks, national forests, and wildlife refuges have wilderness areas within them. For example, Yosemite National Park contains 760,917 acres, of which 677,600 have been designated as a wilderness area. The 83,317 acres outside the wilderness area are allowed to have campgrounds, cabins, lodges, and other facilities.

Wilderness and the Public Lands

In 1803, the United States purchased the Louisiana Territory from France. This territory stretched from the Mississippi River to the Rocky Mountains, and from the Gulf of Mexico to the Canadian Border, for a total of 827,987 square miles. What is now known as the Pacific Northwest was acquired by the United States in 1846 in a treaty with Great Britain, and in 1848 the United States acquired most of the Southwest, including what is now

California, Arizona, Nevada, and Utah, in a treaty with Spain. Most of these lands became known as the *public domain*. Today many people use this term to mean any land owned by the federal government.

The United States federal government encouraged westward settlement in order to cement their hold on the lands and to boost the economy by taking advantage of the vast, untouched resources, including farmland, timber, and minerals. The first land-use agency created by the federal government was the General Land Office, established in 1812 to administer the sale of federal lands to settlers.

The first congressional act relating to the wilderness was the Preemption Act of 1841, which set guidelines for the sale of public land designed to expedite westward settlement. In 1862 Congress passed the Homestead Act, opening all public domain lands to unrestricted settlement.

Not long after these laws opened the floodgates of expansion, the federal government began to see that they needed to retain some control over the western lands. Northern Wyoming was recognized as an area of particular scenic beauty by explorers such as John Colter in the early 1800s and James Bridger in 1830. By the late 1860s some individuals feared that this beauty would fall victim to uncontrolled exploitation. Members of the Washburn Expedition of 1870 called for a national park at Yellowstone to save this natural treasure for the enjoyment of present and future generations. When Congress finally created the park in 1872, they claimed it as "a public park or pleasuring ground for the benefit and enjoyment of the people."[1]

In the meantime, an entire logging industry was being created in the public forests without the consent of the government. In effect these loggers were stealing trees from federal land and selling them illegally. When Carl Schurz was appointed Secretary of the Interior in 1877, he became the first high-ranking public official to seriously try to put a stop to this abuse. His actions were met with scorn, and Congress cut the funds for his agents in the field.

When President Theodore Roosevelt took office in 1901 he ushered in an era of conservation the country had never seen. Roosevelt was the first president to make managing and conserving our natural resources a main priority. In 1905 he created national forests and established the United States Forest Service (USFS), primarily to manage the leasing of national forest lands to the timber industry and to halt illegal logging and grazing. Like their Interior Service predecessors, when USFS rangers first began patrolling the forests to stop abuses and watch for forest fires they were often beaten and shot at by men who wanted to log the forests indiscriminately.

The primary goal of the USFS was, and still is, to best manage the sale of timber from the national forests. Over the years, however, the environmental movement has grown tremendously, and today all of the federal land agencies must balance conflicting objectives. Some people would like to chop down every tree and dam every river, while others would like to save everything in its pristine state. It is the job of the federal land agencies to come to some rational middle ground.

Today the four federal agencies responsible for managing the federal public lands are the Bureau of Land Management (BLM), National Park Service (NPS), United States Fish and Wildlife Service (USFWS), and the United States Forest Service (USFS). The first three fall under the Department of the Interior. The USFS is part of the Department of Agriculture (USDA). The lands these agencies administer make up about 640 million acres—an area covering nearly one-third of the total 2 billion acres of land in the United States. A great portion of this land is in Alaska, with about 240 million acres. The rest is spread out over the contiguous United States and Hawaii, with the majority in the western states.

The public lands are divided among the different agencies according to the federal government's intended use for it. Some is protected in national parks, preserves, or wildlife refuges, but much of it is open to logging, cattle grazing, mining, or recreational use. All of the agencies oversee lands that are designated as wilderness (i.e., wilderness areas) by Congress, and thus fall within the National Wilderness Preservation System.

Natural Resource Lands

Natural Resource Lands are those administered by the BLM. These lands make up 265 million acres, the largest segment of the federal lands. Much of the BLM land is located in the southwestern states and includes many areas that have been considered wasteland historically. These areas, for the most part, do not contain natural wonders, as in the NPS lands, or vast forests, as in the USFS lands. For this reason the BLM lands have often been called "the leftover lands," or "the lands nobody wanted," but these areas do contain places of beauty and natural resources.

Probably the best known administrative duty of the BLM is the leasing of public grazing lands. Around 21,000 ranchers graze cattle, sheep, and goats on 164 million acres of BLM land. Fees charged for grazing are a major controversy in the West. Ranchers are charged far less to graze on public lands than on private, and many people complain that the U.S. government is losing out on important revenue. In 1993, for example, the federal fee per month per head for grazing was $1.86, compared to an average of $10.03 on private lands in the 11 major grazing states. It is the rancher's political clout that is most responsible for keeping the rates down. In 1993, Secretary of the Interior Bruce Babbitt proposed raising the fees to $4.28 per month per head. This proposal drew heavy opposition from ranchers, some of whom threatened to sue the federal government if the new rates went into effect, causing Babbitt to back down.

The BLM also controls leasing for oil, gas, coal, and oil shale mining and drilling sites and mining leases for gold, silver, iron, copper, lead, and other minerals. It manages 7.9 million acres of commercial forest, and 35 million acres of wetlands. Of all the BLM lands, 5,176,165 acres are protected as wilderness areas.

National Park System

The first national park in the world was Yellowstone, established in 1872. The National Park System of the United States, established in 1916, was also the first of its kind. All other national parks in the world, from those in Australia to Zaire, owe their conception to a large degree to the parks in the United States.

Over the years, the National Park System has evolved to include 54 national parks, as well as urban parks, national monuments, national memorials, and even road systems known as parkways. Some say the original vision of the system has been skewed by the introduction of some of these elements, and that the NPS should concentrate on natural parks. Others maintain that even with its broad set of duties, the NPS is the best agency to manage the broad range of parks and facilities under its control.

A *national park* is an area in which the public can enjoy a specific natural, and generally undeveloped wilderness setting. National parks encompass approximately 49.8 million acres. The figures listed here include Death Valley and Joshua Tree National Parks, which were authorized by the California Desert Protection Act in 1994. Similar versions of the bill have been passed by both houses of Congress, though it has not yet been signed into law as this book goes to press.

National monuments are similar to the national parks, but are usually smaller. These areas are set aside to protect natural wonders or historical sites. Devils Tower National Monument, Wyoming, for example, was established to protect a specific natural wonder—the Devils Tower. Declaring a site a monument is often a first step in creating a national park. Today about one-quarter of our national parks were originally national monuments. There are 74 national monuments for a total of 2.2 million acres.

National seashores are protected for their scenic value. These are undeveloped stretches of coastline left in their pristine state.

National preserves are areas set aside to protect specific resources, such as wildlife, oil, coal, or timber. Extraction of oil and minerals is permitted in some cases, on a very limited basis. The national preserves total approximately 23.5 million acres, including Mojave National Preserve, authorized by the California Desert Protection Act of 1994.

National recreation areas are usually located around dams, reservoirs, lakes, or streams, and are meant to provide recreation, such as swimming, boating, or hiking. The national recreation areas contain a total of 3.6 million acres.

The National Wild and Scenic Rivers System and *The National Scenic and Historic Trails System* were both created in 1968. Rivers, once placed in the National Wild and Scenic Rivers System, cannot be dammed, dredged, canalized, or filled, and development on their banks is prohibited. A similar prohibition on development exists for the national trails, though due to their length they occasionally must cross roads and highways. The Wild and Scenic Rivers System protects 7,225 miles of 66 rivers, or about one-fifth of one percent of the total number of miles of rivers in the United States. The National Trails System protects national recreation, scenic and historic trails, including the 2,000-mile Appalachian Trail from Maine to Georgia and the 2,350-mile Pacific Crest Trail through California, Oregon, and Washington. Trails within the National Trails System, however, are under jurisdictions of multiple agencies depending on their locations. Only three national trails are wholly administered as units in the National Park System (see p. 150).

International Historic Site	1
National Battlefield Parks	3
National Battlefield Site	1
National Battlefields	11
National Historical Parks	36
National Historical Sites	72
National Lakeshores	4
National Memorials	26
National Military Parks	9
National Monuments	74
National Parks	53
National Parkways	4
National Preserves	15
National Recreation Areas	18
National Reserves	2
National Rivers	6
National Scenic Trails	3
National Seashores	10
National Wild and Scenic Rivers	9
Parks (other)	11
Total	**368**

Of these units, national parks, preserves, memorials, national sea-shores, and one national river contain wilderness areas. A few recreation areas and national historical parks contain wilderness and are being studied for potential inclusion into the National Wilderness Preservation System.

National Wildlife Refuge System

In 1903, President Theodore Roosevelt set aside Pelican Island, Florida, as a preserve, primarily to protect the island's bird popu-lation. This was the first wildlife refuge in the nation. Over the years, more refuges were designated, and in 1966 Congress created the National Wildlife Refuge System to coordinate all of the refuges. The USFWS, an agency within the Department of the Interior, is responsible for managing the system, which is made up of over 88 million acres; more than the National Park System. All but 13 million of those acres are located in Alaska.

The refuge system consists of 424 wildlife and waterfowl refuges. About three-fourths of these were set up primarily to protect waterfowl habitats, and more than 600 of the 813 bird species in the United States spend some of their time each year within a wildlife refuge. The refuges also contain 220 species of mammals and 260 species of amphibians and reptiles, including 63 endangered species. The lands within this system are managed with the welfare of the wildlife as a first priority, but this does not offer complete protection from development. In order to be wholly protected, an area must also be incorporated into the National Wilderness Preservation System as a wilderness area. Of 88,615,896 acres in the National Wildlife Refuge System, 20,685,327 acres are designated wilderness.

National Forest System

Falling under the jurisdiction of the USDA is the USFS, which manages 191 million acres of national forest that comprises about 18 percent of the remaining commercial forestland in the country.[2]

An agency motto for the USFS is, "Land of Many Uses." The foremost of those uses is still leasing timber rights, but the national forests also provide grazing areas for ranchers, places for recreation, and wilderness preservation.

USFS timber leasing policies have come under harsh criticism in the past decade because of what many private environmental groups claim are "below-cost timber sales." An accounting system called the Timber Sale Program Information Reporting System (TSPIRS) was designed to prevent below-cost sales. The USFS stands by the system, but private groups, like The Wilderness Society, claim that the USFS understates costs and overstates revenues by as much as 200 percent. For example, the USFS charges timber companies for construction of bridges, culverts, and road surface, but not for road base, which accounted for 63 percent of total road costs in 1992. Also, the USFS does not account for subsidies paid to timber companies in their profit report. According to The Wilderness Society, the USFS lost $244 million in timber sales in 1992.

In 1972 and 1977 the USFS conducted what were known as the Roadless Area Review and Evaluations (RARE I and RARE II) in order to determine how much wilderness existed within the national forests. The RARE II survey identified nearly 3,000 potential new wilderness areas for a total of 62 million acres. So far, 34,466,831 acres in the national forests are set aside as wilderness areas.

National Wilderness Preservation System

The lands designated by Congress as wilderness areas combine to form the National Wilderness Preservation System, which includes a total of 103,438,972 acres. When the Wilderness Act was passed it required all four land agencies to examine their holdings and make recommendations as to which areas might qualify as wilderness areas. Most of these tracts were then voted on for inclusion into the National Wilderness Preservation System; some of the areas are so controversial that a vote has not yet been taken.

Just because an agency recommends an area as an addition to the system does not mean that Congress will approve it. Many recommended areas have been caught up in bureaucratic red tape in a legislative system that often considers the issue a low priority. For example, the NPS recommended in 1972 that 2,016,181 acres of Yellowstone National Park be added to the system, but as of 1994, none of this land is designated wilderness. In addition, the political sensitivity of some areas makes adding them to the system difficult. The Arctic National Wildlife Refuge, for example, has 8 million acres of wilderness designated, but an additional 11,285,922 acres within the refuge is not classified as wilderness because there is debate as to which is more important to the good of the nation: preserving the entire area, or extracting the oil deposits which may exist there.

Any lands that are added to the federal agencies in the future may also be reviewed, and additional areas can be reviewed if a member of Congress appeals for a study. Private, state, or local lands can only be added to the system if they are first acquired by the federal government and placed under the jurisdiction of one of the four land agencies. Each individual agency is responsible for managing the wilderness areas within their jurisdictions.

State and Private Wilderness Preservation

While the majority of protected wilderness in the United States is owned by the federal government, many states and private organizations have created parks or preserves of their own to protect wilderness. The state parks include several Redwood state parks in California; Indiana Dunes State Park, Indiana; Ascutney State Park, Vermont; and others stretching across the country. Adirondack and Catskills state parks in New York contain over 6.3 million acres. These two parks were originally created as forest preserves. On establishing these areas, the New York State Legislature wrote, "The lands now or hereafter constituting the Forest Preserve shall be forever kept as wild forest lands. They shall not be sold, nor shall they be leased or taken by any corporation, public or private."[3]

Private organizations are also responsible for preserving large tracts of wilderness. Among these is the National Audubon Society, which maintains several private wildlife sanctuaries across the country, including the 4,000-acre Starr Ranch Sanctuary in Orange County, California. The National Parks and Conservation Association has a $1 million fund for land acquisition. Much of the land purchased by this fund is donated to the NPS for inclusion into the national parks. The Save-The-Redwoods League has spent more than $2 million to help protect redwood forests in California, including the purchase of 120 acres which were donated as an addition to Portola Redwoods State Park.

By far the largest organization devoted to wilderness preservation through private land acquisition is The Nature Conservancy. The conservancy was founded and exists today primarily to purchase land for preservation purposes. The organization owns and manages over 1,300 preserves worth over $400 million. This includes 1.3 million acres across the United States and Canada, plus additional holdings around the world.

Issues and Controversies in the American Wilderness

In the battle between economic and environmental groups, certain issues concerning wilderness preservation stand out as the most controversial and emotional, in which the ideologies of the two groups are completely incompatible but must somehow be reconciled. This section will explain these conflicts and detail the steps being taken to find solutions.

America's Forests

More than any other ecosystem type in the American wilderness, the forest is prone to controversy, primarily for two reasons. First, forests contain some of the last undisturbed natural areas in the country, and many of those areas are exceptionally beautiful and biologically diverse. At the same time, the forests generate billions of dollars for local economies through the timber industry. The question of how much forest should be cut is one that sets environmentalists against logging companies and the areas that rely on those companies for economic survival. The issues involve not only the quantity of wood to be cut, but also the type of forest, logging methods, forest fire management, and the fate of endangered or threatened forest species such as the northern spotted owl.

In the region that now makes up the contiguous United States, forests once covered about 850 million out of the total 2 billion acres. Due to clearing land for farming, development, and logging, by 1952 only 664 million acres of forests remained. Once the technique of planting new seedlings to replace those cut by industry was implemented, the amount of forest lands rebounded considerably.[4] By the early 1990s forests covered an estimated 731 million acres.[5] Of these acres, approximately 431 million acres are classified by the USFS as commercial timberland. This timberland is defined as being "capable of producing at least 20 cubic feet of industrial wood per acre per year and not reserved for uses that are not compatible with timber production."[6] In other words, protected areas, such as national parks and wilderness areas, are not considered timberlands and no timber harvesting is allowed on these lands. National forests, however, are open to logging. Ownership of the timberland in the United States is divided as follows:

Private Ownership (non-forest-industry)	276 million acres
Forest Industry Ownership	71 million acres

National Forests   85 million acres
(public ownership)

Other Public Holdings   51 million acres[7]
(state and local ownership)

Both environmentalists and loggers agree that the increase in total forest during the latter half of this century is a significant gain, and cause to celebrate, but unfortunately the heart of the conflict centers not around the total amount of remaining forest, but around the old-growth forest and clear-cut logging methods. Old-growth are stands of trees that have never been cut, with undisturbed vegetation and wildlife, and trees as old as 1,000 years or more. Redwoods, in fact, can live for over 2,000 years, and sequoias 3,000.

Of an original 25 million acres of old growth in the Pacific Northwest, less than 10 percent remains. One-third of that is protected in parks and preserves. At the current rate of timber cutting, the remaining two-thirds will be gone by the year 2010.[8] These areas contain the largest percentage of commercially salable wood. Much of the other forest area has been logged in the last 50 years or so, and many of the new trees are not yet mature enough to be cut profitably. The timber industry desperately wants to get at the remaining unprotected old growth, and the environmentalists want to save it.

If old-growth forests continue to be cut, naturalists foresee the end of an entire type of ecosystem, and many of the plants and animals that live there. If these forests are put off-limits, the forest industry foresees the loss of thousands of jobs, with a devastating effect on the economy of the Pacific Northwest. In either case, this is probably the most publicized and emotionally fired environmental battle in the United States in the latter half of this century. The two groups predict either the loss of a natural treasure, or the loss of a way of life for thousands of loggers, with no easy compromise.

Most of the virgin forest remaining in the country is located in the Pacific Northwest. This is also the country's most prolific logging area and the center of the logging controversy. Most other forests in the United States have either been cleared or burned by humans in the last three centuries. These young or replanted forests are not valued as much by naturalists or logging companies, because they contain less biological diversity and do not contain as much wood as the old-growth forests. In fact, due to the sheer immensity of the old-growth trees, these forests contain the greatest biomass of any forests in the world; up to ten times more biological matter (including primarily wood, but also plants and shrubs) than the forests planted to replace them.[9]

In 1989 the issue became much more complicated with the discovery that logging was threatening an endangered species, the northern spotted owl. Passed by Congress in 1973, the Endangered Species Act makes it a federal crime to "capture, harass, or kill," any species on either the "endangered" or "most threatened" species lists.[10] An *endangered species* is one that is likely to go extinct in the near future if some action is not taken to save it. A *most threatened* species is one that is likely to become endangered in the near future.

The USFWS and the National Marine Fisheries Service are responsible for administering the act, and it is these agencies that conduct studies and ultimately decide which species to place on the lists. Currently there are 1,140 species on the endangered list worldwide, with 618 of those native to the United States. Anyone can petition the USFWS to study a particular species and potentially add it to one of the lists. Environmentalists have taken full advantage of this law, often petitioning to list a species in order not only to save that species, but to preserve an entire ecosystem from development.

In the spring of 1989 logging opponents found a species that would help them preserve a good portion of the remaining old-growth forests. A coalition of organizations headed by the Seattle Audubon Society petitioned the USFWS to grant the northern spotted owl endangered status. Timber sales from most national forests in western Oregon and Washington were almost immediately blocked until studies could be undertaken. The spotted owl can only survive in the complex ecosystem of Pacific Coast old-growth forests. If all or most of the remaining old-growth forest is cut, the last of the northern spotted owls will perish.

The timber industry was furious at this development. Logging supporters wanted to overturn the Endangered Species Act completely. Senator Robert Packwood (R-OR) estimated that 100,000 workers in the timber and timber-dependent industries could lose their jobs to save the spotted owl. The American Forest Resource Alliance, a pro-timber organization, said that, "the economic impact of preserving the owl and its habitat could mean a loss of 2.6 billion dollars for affected communities over the next ten years."[11]

Environmentalists say that if jobs are saved today by allowing logging in old-growth forests, those jobs will still disappear once the old-growth is all cut down. They feel it is better to lose some jobs now and save the virgin forests than to lose both the virgin forests and the jobs in 15 years.

In 1989 the U.S. government gave six wildlife biologists the assignment to come up with a plan to balance the owl's survival with the needs of the timber industry. The team, led by Jack Ward Thomas, came up with a plan to set aside just over 8 million acres as protected owl preserves. Each preserve would theoretically support 15 to 20 owl pairs. This plan was found to be unacceptable to the timber interests, who demanded more acreage for their use.

In 1993 President Bill Clinton and Vice President Al Gore met with both sides at a "timber summit" in Portland, Oregon, and the president promised a plan to be issued within 60 days. Clinton set up three teams of scientists and economists to study the situation. One of those teams, the Forest Ecosystem Management Assessment Team (FEMAT), came up with ten possible options, ranging from halting all old-growth logging, to continuing a substantial amount of cutting. After studying these options, President Clinton chose a plan that will allow 1.2 billion board feet to be cut each year, down from a high of 5 billion board feet per year that was cut in the mid-1980s. FEMAT economists estimate that this will result in a net loss of around 10,000 jobs. Logging officials put their estimate at closer to 85,000 jobs. In order to soften any economic impacts, the Clinton plan includes a $1.2 billion aid package to help diversify the economies of local communities. Also included are incentives for timber companies to process logs in the United States instead of sending raw logs overseas, in order to create more sawmill jobs.

As for saving the spotted owl and other forest species, the plan centers around protecting watershed areas, theorizing that these areas are the core of the ecosystem. The survival of salmon and other fish are taken into strong consideration as well. Logging practices have threatened fish populations in the Pacific Northwest, as well as the fishing industry, which employs over 60,000 people.

When this plan was announced in July 1993, it drew strong criticism from both sides. The timber industry still cried that too many jobs would be lost and the environmentalists still complained that no more old-growth forests should be cut. Either way, the days of unlimited logging are over, but with sound management, a constant supply of wood can always come out of the region. Even though neither side is happy about it, Clinton's plan is the best chance so far to salvage some old-growth forests and the timber industry.

For those areas that are allowed to be logged, another controversy concerns the method by which they are cut. The method most favored by the timber industry is that of clear cutting. This is the most productive and economically viable way to cut trees. It also results in the quickest regrowth of timber, but several factors make clear cutting unappealing to the general public. The technique itself involves cutting every standing tree in a specific area. The remaining undergrowth and debris is burned off to simulate the natural fire process. This allows trees to regenerate more easily, without having to compete for sun with undergrowth, but it also destroys the diversity of the forest. New pine trees are planted, but other forms of plant life are not, and shade from the planted trees actually discourages the growth of other species, as well as younger trees. The variety of plant species found in an old-growth forest does not exist in a second generation forest after clear cutting. For this reason, many species, including many small mammals and the spotted owl, cannot survive in a second generation forest.

Another problem with clear cutting is that without the root systems of the undergrowth, runoff from clear-cut areas washes away top soil and silt that can choke much of the life out of surrounding rivers and streams. Several once-prolific salmon runs have been devastated and others are in danger. The loss of salmon pits not only environmentalists, but salmon fishermen as well against the timber industry. In this case it is fishermen who are losing their jobs, and as salmon stocks continue to shrink, the fishermen's way of life is also in serious danger.

A big problem with clear cutting from a public perspective is the fact that it scars the landscape, if only temporarily. People visiting the national forests are often horrified to see vast tracts of charred tree stumps. To counteract this problem, the USFS began a policy to limit the aesthetic damage by drawing the lines of areas to be cut in a way to better blend into the natural layout of the forest. They also began to leave buffer zones along roads to block the view of clear-cut areas from passing motorists.

Many environmental advocates are not satisfied with simply trying to hide the destruction from view. Some say that loggers should not be allowed to clear cut at all, but should cut selectively, leaving a certain number of trees standing in any given logging area. Opponents claim that under this method the best trees will be cut, leaving a second-rate forest with undergrowth that discourages new tree growth.

One area in which the conflicting views of environmentalists and loggers came to head in the early 1990s was the Nez Perce National Forest in Idaho. This area is the largest roadless wilderness in the contiguous United States. The national forest contains 2,258,542 acres. A total of 936,510 of those acres are protected as wilderness areas, but the USFS has approved limited logging and the construction of access roads into the unprotected areas.

This forest became a focus of controversy over both forest management and radical environmental tactics. During the summer of 1992 about 25 members of an environmental movement called Earth First! camped out in the Nez Perce and began a campaign to stop the logging. The group uses such nonviolent tactics as lying down across roads to prevent access, sitting in trees, and in one case a member buried himself in a road up to his chest. By the following summer the camp swelled to 40 people, mostly college students, and included 2 converted school buses in which some members planned to stay indefinitely.

Earth First! is infamous among traditional environmental groups because they have historically advocated monkeywrenching, which includes such techniques as vandalizing construction equipment and drilling spikes into trees to make them unsafe for cutting. Similar acts were performed in the Nez Perce area, and by the end of the summer of 1993, tensions between local residents and the Earth First! members were at a dangerous level. In one case a lone Earth First! member was ambushed and beaten in the forest.

The USFS maintains that the areas approved for logging are not essential to the survival of the ecosystem, and that the protests will delay the cutting but will not stop it. Earth First! members vowed to keep up their protest until logging in the Nez Perce is halted. The dispute provides an example of the anger, emotion, and broad difference of opinion common to many wilderness conflicts.

Management of wild fires in the forests is also an issue that creates some controversy. Forest fires have occurred since the first forest evolved on earth, and man has manipulated the forest fire in one way or another since early civilization. When white settlers arrived in North America, they found that the native tribes routinely burned areas of forest. Grass quickly grew in the burned areas, attracting game and creating fertile hunting grounds. The white settlers soon adopted the use of fire to clear land for settlement. This technique was used regularly for the next three centuries. Without a means to put the fires out, however, they often spread out of control. Several fires burned between 3 to 5 million acres each in the late 1700s and early 1800s. In 1780 a member of the Connecticut state legislature described a day darkened by smoke in which, "None could see to read or write in the house, or even at a window, or distinguish persons at a small distance."[12] The practice of clearing land with fire slowly died out as the nation's population grew. The large amounts of lumber in the forests became more valuable, and the uncontrollable fires became too dangerous with homesteaders throughout the land.

In the twentieth century, with the advent of national forests and national parks, a new policy of extinguishing fires came into effect. Teams of USFS fire fighters were formed to put out all fires immediately in order to save the parks and forests from destruction. By 1935 the USFS was so intent at squelching fires they initiated a "10 A.M. policy" that called for all fires in the national forests to be extinguished by 10 A.M. the morning after they started.

By the mid-1960s scientists and foresters alike started to theorize that suppressing all fires was not such a good idea. Since many areas of forest had not burned in 100 years or more, the level of fuel on the forest floor was becoming dangerously high. Also, fire clears the forest of debris and provides nutrients for new growth to begin. Studies have shown that forests have a remarkable ability to regenerate themselves after a fire. Several varieties of tree, such as the sequoia, can survive lower intensity fires, and others have built-in mechanisms to reproduce after a fire. The lodgepole pine, for example, has two types of cones. One is the nonseratinous, which opens and distributes seeds every year. The other is the seratinous, which will only open after a fire melts away the resin that holds the cone together. These cones can stay on the tree for up to 25 years waiting for a fire to release their seeds. Studies have shown lodgepole pine seed densities of 50,000 to 1 million per acre after a fire.[13] One pair of researchers, Jay and Phyllis Anderson, found up to 79 seedlings per square meter a year after a fire in Yosemite.[14] Eventually the strongest seedlings survive, leaving an average of 325 trees per acre in a mature forest.

The clearing of dead debris is possibly the forest fire's greatest benefit. If smaller fires do not occur on a regular basis in the forests, the dead and dying brush and trees build up to dangerous levels, creating fuel for larger, more destructive fires. In 1972 the NPS officially recognized this fact and started a "let burn" policy for naturally occurring fires within the national parks; the USFS followed with a similar policy for their wilderness lands. These plans state that lightning fires should be allowed to burn naturally unless they threaten people, historical or cultural sites, special natural features, threatened or endangered wildlife, or move outside park or wilderness area boundaries. All accidental man-made fires are to be fully suppressed immediately. Finally, the plans call for controlled burning in certain cases to clear out areas with an excessive buildup of fuels.

Many forest fires are started accidentally by man, through discarded cigarettes or untended campfires, but lightning also starts a large number of fires each year. In fact, 500,000 cloud to ground lightning strikes occur over the world's forests every day. About 10 percent of these, or 5,000, result in fires.[15]

The "let burn" policy created controversy from its inception, but in the summer of 1988 the debate came to a head when a massive series of fires blazed across Yellowstone National Park and surrounding parts of Wyoming and Montana. The fires raged out of control from mid-June until mid-October, destroying over 1 million acres of forest before they were finally extinguished. The sheer enormity of the fires of 1988, and the fact that they were centered in Yellowstone, a crown jewel in the park system, intensified the debate on forest fire management. Some of the fires that started in June were allowed to burn for a period before crews were sent in to fight them. Some angry local residents claimed that the fires could have been put out immediately but were allowed to grow to unstoppable proportions. Many fingers pointed at the NPS, claiming that their "let burn" policy was responsible for the conflagration.

It is unclear what the result would have been had the NPS attacked all fires immediately that summer, but it is important to note that 1988 was the driest summer on record in Yellowstone, with practically no rain falling in June, July, or August. Also, the largest and most costly fire to put out, the North Fork Fire, was started by a group of men smoking cigarettes. This fire was fought hard from the very first day, though it burned over 500,000 acres and cost $25 million to extinguish.[16]

The NPS contends that it was the fire suppression policies of the past that were ultimately responsible for the 1988 fires, not the "let burn" policy. As Park Ranger Don Buss said, "The last time the park burned like this was in the 1600s. All this fuel has been building up ever since, and we've added to it by putting out all fires for over 100 years."[17] Under the "let burn" policy, between 1972 and 1987 fires totaling only 34,157 acres were allowed to burn, but this wasn't enough to clear out a century of accumulated fuel. Today, the "let burn" policy remains in effect. The forests that burned in Yellowstone are recovering quickly, and with the heavy underbrush cleared they are considered to be virtually fire proof for at least a century.

America's Rivers and Lakes

Rivers and lakes in the United States face a variety of threats from several different types of pollutants including sewage, pesticides, acid rain, herbicides, and other chemicals. Many of America's rivers flow through both wild and urban areas, picking up contaminants as they go. In 1989 the Environmental Protection Agency (EPA) released a report that said 17,365 rivers, streams, and bays in the United States suffer from these forms of pollution.[18]

The problem of contaminants in rivers is one which varies dramatically from region to region. Rivers in the eastern United States and the Midwest tend to have greater concentrations of pollutants because they are closer to industrial centers. In 1969, the Cuyahoga River near Cleveland, Ohio, was so polluted that it actually caught on fire. The Federal Water Pollution Control Act of 1972 (Clean Water Act) was designed to set standards and guidelines for industrial polluters in an attempt to clean up the rivers. Much progress was made over the next 20 years, but in 1993 Congress decided to get even tougher on big industry, passing an amended Clean Water Act with stricter standards.

An issue that has even more bearing on wilderness areas is the building of hydroelectric dams. These dams create terrific amounts of hydroelectricity, but cause problems to the wilderness at the same time. The construction of dams has flooded some truly spectacular wilderness, such as the Hetch Hetchy Valley in Yosemite National Park. John Muir fought vigorously against this project, and considered the Hetch Hetchy Valley one of the most beautiful areas of the park, but he lost his battle when the dam was approved in 1913.

Dams also make it difficult, or even impossible, for fish to swim upstream. This is especially prevalent in the case of salmon in the Pacific Northwest, which need to migrate upstream in order to reproduce. Fish ladders have been built around the sides of many dams, which vary in construction, but usually involve a section where water from the river or stream bypasses the dam and flows down a set of stairs, allowing the fish to swim up and over the top. These can be difficult for the salmon to overcome. Some dams have no ladders at all, and fish must be captured below the dam, loaded in a truck, and driven around to be released at the top. Since schools of salmon always return to the streams in which they hatched, the fish that breed in these streams are entirely dependent on these trucks for the very survival of their lineage. In March 1994, Secretary of the Interior Bruce Babbitt suggested the destruction of two dams in Olympic National Park to help the salmon population. The plan created some controversy, however, and has yet to be approved. In April of the same year, the Pacific Fishery Management Council banned salmon fishing off the coast of Washington State, and placed strict limits on commercial and recreational salmon fishing off California and Oregon.

Other factors have contributed to a decline in the salmon population as well, including pollution and erosion of soils due to clear-cut logging operations. After a tract of forest has been clear-cut, no vegetation is left to hold the soil, and it can wash away with rain, choking streams and killing fish. In the 1880s an estimated 10 to 20 million salmon returned from the Pacific to spawn in the rivers and streams. By the early 1990s that number was down to 2.5 million.[19] Several species of salmon are near to being listed on the most threatened species list, including coho, sockeye, and chinook.

One type of pollution that can threaten fish of all types, even in the most remote locations, is acid rain. Coal and oil combustion from factories and automobiles release gasses into the air, including sulfur dioxide (SO_2), nitric oxide (NO), and nitrogen dioxide (NO_2). When these gases attach themselves to water molecules they fall to the ground as acid deposition in the form of acid precipitation: rain, snow, sleet, hail, dew, fog, or mist.

In the northeastern United States, where acid rain is a serious problem, the average rain today is nearly ten times more acidic than normal. In cities this can cause accelerated rusting of metals and decay of buildings and materials. In the wilderness it can cause contamination of the soil and water. Acidity in lakes often results in death of aquatic life, ranging from younger fish in a lightly acidic lake, to every living thing in a heavily acidified lake. This in turn affects waterfowl and other species that depend on the lake for food.

Acid rain is one of the more frustrating threats to our wilderness today, because the problem originates so far away from the damage. Someone in a city is probably unaware that the exhaust from their car is contributing to the death of a once-pristine lake in the wilderness, and in many cases this is exactly what is happening. Exhaust and fumes from urban areas can travel long distances in the atmosphere and clouds before causing damage in remote areas.

Acidity is measured on what is known as the pH scale. This means the potential of hydrogen, or the ratio of positively charged hydrogen ions to negatively charged hydroxyl ions. The scale goes from 0 to 14. At 7 a substance is neutral, with an equal number of positive and negative ions. The larger the number, the more alkaline; the smaller the number, the more acidic. The scale is logarithmic, so that at each step a substance is ten times as alkaline or acidic as the previous level. With a pH of 2, for example, lemon juice is ten times as acidic as vinegar, with a pH of 3.

Normal, or "pure" rainfall today has a pH of 5.6, or slightly acidic. The lowest pH ever recorded in rain, due to fossil fuel pollutants, was 2.2 in Wheeling, West Virginia. This is about eight times more acidic than vinegar, and nearly as acidic as lemon juice.

While scientists agree that acidity as found in acid deposition is toxic to many species of plant and animal life, the actual extent of the problem is still a source of debate in the scientific community. Some scientists point to the large number of fishless lakes in the Northeast and blame acid rain, while others claim that there are other factors involved as well. Many fishless lakes existed, for example, before the advent of acid rain. Studies have shown, though, that when pristine lakes are acidified in test cases, plant and animal species do die. Most scientists agree that there is a definite negative effect, even if they disagree on the magnitude, and with the passage of the Clean Air Act of 1990, Congress committed the U.S. government to address the problem.

It is possible to counteract the effects of acid rain in a lake by adding lime (calcium oxide) which has a pH of about 12.5. The alkalinity of this substance can offset acidity and move the pH toward neutral, but this method is costly and only a temporary solution. The only real, long-term solution to the problem of acid rain is to cut down fossil fuel emissions. The Clean Air Act of 1990 requires the 111 largest coal-burning electric power plants in the nation to meet strict new air-quality standards through such techniques as using coal that has a lower sulfur content, or new sulphur-cleansing technology in their smokestacks. The act also calls for lower auto emissions and includes a goal of cutting sulphur emissions by 10 million tons a year by the year 2000 and cutting nitrogen oxide emissions in half.

America's Prairies

When pioneers first arrived to settle the Great Plains, there existed 200,000 square miles of prairie, comprised of three types: shortgrass, mixed, and tallgrass. The tallgrass prairie originally covered 140 million acres. Today, according to The Nature Conservancy, only one hundredth of one percent of the tallgrass prairie remains. The other types of prairie have disappeared in similar fashion. American farms on these lands now feed our country and much of the world, with $200 billion in crops harvested every year.[20]

The prairie farmlands are much more fragile than many people realize. During the 1930s, poor farming practices, coupled with a drought, caused millions of tons of topsoil to blow away in what became known as the Dust Bowl. Today, poor farming practices continue to deplete the topsoil. Also, much of the water for irrigation and drinking comes from a huge underground reserve stretching from Nebraska to Texas called the Ogallala Aquifer. Every year farmers pump 6 million acre-feet of water from the aquifer, though nature only adds 185,000 acre-feet per year. If this practice continues the aquifer will be depleted sometime in the twenty-first century, leaving many farms with no source of water.[21]

The remaining native prairies in the United States have little wilderness protection. Badlands and Wind Cave national parks cover 271,051 acres in South Dakota. Theodore Roosevelt National Park encompasses 70,447 acres in North Dakota. Environmentalists have argued for a national park on 320,000 acres in the Flint Hills of Kansas, but no real progress has yet been made.

National wildlife refuges are located throughout the plains states to protect the regions' wetlands and waterfowl. The majority of the wildlife refuges are in the far north. Kansas, for example, has only 4 refuges with a total of 56,279 acres, while North Dakota has 63 refuges (more than any other state) covering 290,216 acres. Migratory birds travel north and south from Canada to Mexico through the plains states on what is known as the central flyway. The flyway is particularly narrow in Nebraska, where the birds congregate in the Platte River Valley. Sandhill cranes spend almost a month in the valley during their northern migration. Other species include endangered whooping cranes, piping plovers, and least terns. Today, nearly 75 percent of the original wetlands in the valley have either been drained or have dried up due to the shrinking water table.[22] The Platte River itself is also diverted to farms, causing parts of the river to completely dry up during the irrigation season. Even with these impediments however, the sandhill cranes are still thriving, with a population of approximately 500,000.[23]

The Nature Conservancy has done the best job of preserving the prairie grasslands, through the purchase of 551,000 acres, including the 6,000-acre Cross Ranch Preserve in North Dakota, the 7,200-acre Konza Prairie Preserve in Kansas (managed by Kansas State University), the 54,000-acre Niobrara Valley Preserve in Nebraska, and the 35,000-acre Tallgrass Prairie Preserve in Oklahoma. To return the ecosystem to its original form, the Conservancy works to encourage the growth of native plant species and has released herds of bison on several preserves, including 300 head on the Tallgrass Preserve.

To many people the bison, better known as American buffalo, represent the destruction of the American wilderness. When white settlers first ventured onto the plains, the land was thick with bison. Estimates of the original bison population range from 25 to 60 million head. White hunters on horses and in trains decimated the herds for their hides, or just for sport. By the turn of the century, less than 100 years later, fewer than 600 bison were still alive. Through captive breeding programs, the population has rebounded to 120,000 today.[24] Most of these live either on preserves, like those run by The Nature Conservancy, or on ranches where they are raised for their meat. The last "free-ranging" herd consists of 3,000 head living in Yellowstone National Park, but even these are not wholly free. Any of the animals that wander out of the park's boundaries are shot by park officials.

America's Deserts

The American deserts have not been settled as extensively as other regions in the United States due to the lack of water necessary for agriculture. For this reason, the great deserts of the American West contain some of the most isolated wilderness in the country today, but even these often inhospitable regions face growing pressure from human populations. The deserts host a myriad of unique plant and animal species, but the ecosystem is a fragile one that does not rebound easily when damaged, and mining operations, off-road vehicles, military reservations, and irrigation have taken their toll.

Four major deserts exist in the United States, including the Mojave in California, Nevada, Arizona, and Utah; the Chihuahuan in both Mexico and New Mexico; the Sonoran in Mexico, California, and Arizona; and the 160,000-square-mile Great Basin in Nevada, Oregon, Idaho, Wyoming, Arizona, and California. The Great Basin is actually made up of 150 basins, with 160 separate mountain ranges.

To date, efforts to preserve wilderness in the deserts have concentrated mainly on the spectacular scenic areas, of which there are many. National parks are located at such well-known sites as the Grand Canyon and Petrified Forest in Arizona; Big Bend, Texas; and Bryce Canyon, Capitol Reef, Arches, and Canyonlands in Utah. These parks provide a great service in saving specific scenic wonders, but in many ways they are quickly becoming islands surrounded by a sea of degraded desert. Military airplanes use the deserts as bombing ranges and tanks practice maneuvers. The growing human population diverts water for both domestic and agricultural use, lowering the water table. Chemicals from mining operations leach into that same water table. Grazing cattle cause erosion and trample vegetation along streams, and off-road recreation vehicles carve crisscross patterns across the landscape.

One of the biggest environmental problems to develop in Utah in recent years is a huge surge in mountain bike riding. The number of mountain bikes in the United States grew from 250,000 in 1982 to 25 million in 1994, and the desert surrounding Moab, Utah, became known as a mecca for riders.[25] The sheer numbers of riders leave rutted trails, which in turn cause significant erosion during rains.

The Sonoran Desert, further to the south, is beset by a completely different problem. Over 2,000 plant species (nearly 10 percent of all plant species in the United States) are found in the Sonoran Desert.[26] This area is the only desert in the United States that has trees as the major component of the environment. Ironwood and mesquite trees contribute an important function for developing saguaro and other cactus species. These cactus live up to 500 years, but require shade provided by the trees for their first 50 or so years. Several bird species and reptiles are also dependent on the trees, as are the endangered bighorn sheep and Sonoran pronghorn antelope. Only 150 of the antelope are still alive in the United States, with about 350 on the Mexican side of the border.[27]

Mesquite wood has become so popular for use in barbecues that the trees are fast disappearing in many areas, as they are cut and sold. This problem is more prevalent on the Mexican side of the border, but as the trees are becoming scarce, they are being cut farther and farther north. Ironwood trees are also cut for firewood. In one portion of Organ Pipe Cactus National Monument, along the United States–Mexico border in Arizona, more than 40 percent of the mesquite and ironwood trees within 500 feet of the border have been cut or damaged.[28]

In California the desert accounts for 25.5 million acres. On this land there are nine major military bases and 36,000 miles of roads and dirt tracks, compared to 44,000 miles of highways in the entire United States.[29] The California desert contains around 2,000 different species of plants and 600 animals, including the endangered desert pupfish. This tiny saltwater fish is only found in one place, called Soda Springs, and is believed to be virtually unchanged since the Pleistocene epoch 25,000 to 50,000 years ago. Another species of interest is the creosote ring, which is thought to be the earth's oldest living thing. Each individual plant can live 10,000 to 12,000 years.

Legislators are beginning to realize that even desert areas without spectacular scenic splendor need protection. Both the House of Representatives and the Senate have passed similar versions of the California Desert Protection Act, which is expected to be signed into law shortly after this book is published. The act authorizes the creation of three new national parks in the California desert: Joshua Tree, Death Valley, and Mojave. It also creates 76 new wilderness areas for a total of 7,678,895 acres.

America's Wetlands

Aside from old-growth forests, perhaps no natural lands are as prone to controversy as wetlands. Wetlands are areas covered with water at least part of each year, such as swamps, bogs, and marshes. To many farmers and developers they represent valuable space, yet these areas constitute ecosystems important to many species, including large numbers of migratory birds. According to a study by the National Wildlife Federation, 43 percent of all plants and animals on the Endangered and Threatened Species lists rely on wetlands at some point during their lifetimes, even though wetlands account for less than 5 percent of the land area of the 48 contiguous United States.[30]

At the same time, wetlands are disappearing at a phenomenal rate, estimated at 200,000 to 300,000 acres each year.[31] In the past 200 years the continental United States has lost wetlands at a rate of more than 60 acres per hour. Overall, more than 50 percent of the original wetlands in the continental United States have already disappeared. This percentage is much higher in some states, with California topping the list at 91 percent.[32] The majority of this loss can be attributed to lands drained for agricultural use, with land development ranking second. Many wetlands today are owned by farmers or developers who feel that the government has no right to tell them how to manage their lands.

It is for reasons such as these that wetland protection has become a hotly debated issue. Environmentalists want to save as much wetland as they can, while many farmers and developers insist that their individual projects will leave plenty of remaining wetlands, and that environmental regulations which hamper business are unfair and bad for the economy. The difference of opinion between these two groups is impossible to bridge in most cases. Development destroys wetlands, but wetland protection has an economic price, and neither side is willing to give in without a struggle.

Ironically, all wetlands in the United States are already officially protected, theoretically. The Federal Water Pollution Control Act of 1972 (known as the Clean Water Act) and additional regulations issued under Supreme Court order by the Army Corps of Engineers in 1977 require a permit from the Army Corps of Engineers before anyone can add any fill material to a wetland. The Corps reviews all applications, checking for environmental impact, and holds a public hearing. A permit cannot be approved if "discharge will substantially damage the aquatic ecosystem; if practicable alternatives exist that would have less adverse impact on the wetland; if the discharge violates other environmental standards, such as those dealing with state water quality, endangered or threatened species, or discharge of toxic chemicals; or if all practicable steps have not been taken to minimize the impact on the wetland."[33]

In order to get around these regulations, some developers dug ditches to drain wetlands. The Corps then decided that since these areas were no longer wetlands they no longer had regulatory authority, and the developers were in the clear. The National Wildlife Federation and the North Carolina Wildlife Federation sued the Corps, and to settle the issue, the Clinton administration issued a new regulation in 1993 to clarify that it is illegal to drain wetlands as well.[34]

The biggest part of the dispute relates to the legal definition of wetlands, which has yet to be clearly stated. All wetlands may be protected, but a narrow definition would protect much less acreage than a broad one. So far the debate has led to a number of definitions, going back and forth between the two sides, but as of this printing, no conclusive definition has been adopted by the government.

The basic definition of a wetland that was adopted by the Army Corps of Engineers states that wetlands are

> areas that are inundated or saturated by surface or ground water at a frequency and duration sufficient to support, and that under normal conditions do support, a prevalence of vegetation typically adapted for life in saturated soil conditions.[35]

In 1987, *Federal Manual For Identifying and Delineating Jurisdictional Wetlands* was released, but not all federal agencies adopted it. The major sticking point is that it is based to a large degree on how many days an area must be inundated with water each year to qualify as a wetland.

In August 1991, President George Bush, with a policy of "no net loss" of wetlands in the contiguous United States, released a proposed revision of the 1987 manual for use by all federal agencies requiring 15 consecutive days of flooding or 21 days of saturation to the surface during the growing season. According to scientists and government officials, between 30 and 80 percent of lands protected by previous versions of the manual would no longer be considered wetlands. This proposed version was greeted with outrage, and by 22 January 1992, more than 80,000 formal complaints were sent to the EPA.[36]

According to scientists, because wetlands in different types of environments are so diverse from one another, many other factors must be taken into account, and a definition based on the number of days of saturation is not good enough. For example, it is difficult to classify a wetland in a dry area like Arizona under the same strict guidelines that are used in Florida.

When President Clinton entered office in 1993 he placed the Soil Conservation Corps in charge of administering his wetlands policy. He also reversed a Bush proposal that allowed for development on 1.7 million acres of wetlands in Alaska and called on the National Academy of Sciences to issue a comprehensive definition of wetlands.[37]

Alaskan Wilderness

The last immense tracts of wilderness that exist in the United States today are located in Alaska. The federal government owns 248 million out of a total 365 million acres of land in Alaska, and most of that is wilderness, including Wrangell–St. Elias National Park and Preserve, with 8.7 million acres of protected wilderness, Noatak National Preserve, with 5.8 million acres, and the Arctic National Wildlife Refuge (ANWR), with 8 million. All told, Alaska has 57,444,354 acres of congressionally designated wilderness, compared to 46,114,278 acres for the entire rest of the United States.

The immensity of the wilderness areas in Alaska is mirrored by the size of the conflicts they can create. With nearly 68 percent of the total land controlled by the federal government, local Alaskans complain that they have too little say in how their state is run. Instead, it is distant congressmen in Washington, D.C., who make most of the significant decisions. Controversies include how much wilderness to designate, how much timber to cut, development of oil, gas, and mineral reserves, and animal conservation.

One issue to receive particular attention is the state-sponsored hunting of wolves. During the 1970s and early 1980s the Alaska Department of Fish and Game routinely used helicopters to kill entire packs of wolves at a time in order to maintain large caribou herds for human hunters. After the practice was halted, the local caribou herd near Fairbanks dropped in number from 10,700 to 4,000 head between 1987 and 1991. Caribou hunting was banned and plans were set into motion to reinstate wolf hunts beginning in the winter of 1993. Environmentalists were so upset, however, that the state was hit by a barrage of protests and calls for a tourism boycott. Vacation bookings dropped dramatically, forcing

Governor Walter J. Hickel to call off the aerial wolf hunting. Environmentalists rejoiced, while hunters seethed.

During the winter of 1994, the state of Alaska began a new plan to eradicate wolves. Instead of killing whole packs by air, they began setting traps to kill up to 150 wolves, one at a time, in a limited area. Again environmentalists fumed, and began talking of another boycott. Wolves have such an advanced, close-knit social structure that animal rights groups claim that killing them one at a time is more cruel than shooting an entire pack. Either way, this debate is far from over, as hunters continue to demand larger caribou herds and environmentalists continue to fight for the wolves.

An even larger controversy in Alaska concerns the Arctic National Wildlife Refuge (ANWR). In 1980 Congress passed the Alaska National Interest Lands Conservation Act (ANILCA). Among other things, this act set aside 19,049,236 acres in the far northeastern corner of the state as the ANWR in order to protect one of the last completely untouched ecosystems in the world. This region contains a wealth of wildlife, including moose, wolves, musk oxen, polar, black, and grizzly bears, lemmings, arctic fox, snow geese, and Dall sheep. It is also home to a herd of caribou, known as the Porcupine Herd, that numbers 200,000 head.

Of the 19 million acres in the ANWR, however, only 8 million were given official wilderness status. This leaves the rest of the refuge open to development, upon congressional approval. Among the areas without protection is the coastal plain, which abuts the Beaufort Sea. This area, defined in the ANILCA section 1002(b) and known as the "1002 study area" is believed to possibly contain massive oil reserves. Congress set aside this area for study and so far the issue is still under heated debate.

The U.S. Department of the Interior was given the task of researching the 1002 area for Congress, and in 1987 it issued a final Environmental Impact Statement on the region. This report estimated that the area has a 19 percent chance of containing economically recoverable amounts of oil, and that if recoverable oil does exist, there would likely be a mean average of 3.2 billion barrels. Upon completion of this report, then Secretary of the Interior Donald Hodel recommended leasing the area to the oil companies.

Three-point-two billion barrels is enough to supply all of the oil needs of the United States for 200 days under current consumption rates, but this amount is only a wide estimate of what the ANWR might hold. In reality it could contain anywhere from little or no oil to 9 billion barrels or more, as do the nearby Prudhoe Bay oilfields. In a 1989 report, the U.S. Congress Office of Technology Assessment stated that they believe the 19 percent figure is low and that "there is a higher chance that recoverable quantities of oil exist."[38]

The only way to find out how much oil really does exist in the ANWR is to set up exploratory wells. Congress has not yet decided whether to allow for even exploratory drilling. Dozens of bills have been introduced by both pro-environment and pro-oil groups in an attempt to end the dispute, but so far none have received the necessary support to pass. The debate centers around how important the potential oil might be to our economy, and whether or not drilling would destroy the wilderness qualities of the area.

The coastal plain is the prime calving ground for the Porcupine caribou herd. Every summer the herd moves into the area to graze and bear their young. Large scale drilling operations would most likely have a negative impact on the caribou and other wildlife populations. Oil companies say the impact would be minimal, while environmentalists paint a more destructive picture.

Drilling operations would entail building wells, pumps, and living and storage structures, as well as pipelines and roads totaling about 7,000 acres with an environmental impact on approximately 150,000 acres.[39] The pipelines could block the caribou from their migrations. Oil could leak into the area's water systems, and if tankers transport the oil, there is the potential of damaging oil spills, such as the Exxon Valdez spill, which dumped 250,000 barrels of oil and caused a considerable amount of environmental damage to the southern coast of Alaska in 1989. A spill in the harsher arctic climate could be impossible to clean up, especially in winter.

At the present time Congress does not appear to be in any hurry to come to a decision on this issue. The oil, if it does exist, is not going anywhere. In fact, it most likely will be worth more as time goes by. If another oil crisis were to develop, however, the environmentalists could be hard pressed to stop the drilling of the area.

The Future of Wilderness in America

As the population of the United States continues to grow and the resources become more and more scarce, the remaining areas of wilderness in this country will continue to shrink. This is inevitable. As this remaining wilderness shrinks, those areas that are left will become more and more valuable to environmentalists and developers alike. For this reason, conflicts over wilderness management will continue, and possibly intensify. There is still plenty of unprotected federal wilderness for developers and environmentalists to argue over, but as these areas become either protected or developed, the wilderness in America will be found in smaller and smaller pockets.

It has become increasingly clear in recent years that economics is the basis of every preservation dispute. To set aside any area means to deprive someone of a source of income, whether through timber harvesting, mining, grazing, farming, etc. Following the lead of The Nature Conservancy, environmental organizations are beginning to realize that the easiest way to preserve an area without a battle is to simply buy the area and then not let anyone alter it. Even federally owned public lands can be protected this way with land-swap deals. Many of the larger, wealthier environmental organizations have taken to this strategy in recent years, and it seems to be a trend that will increase considerably in the next few decades.

The trend toward purchasing sensitive areas for protection also points to a change in preservation ideology. Originally it was only the nation's scenic gems, such as Yosemite and Yellowstone, that were thought to need protection. Today it is vanishing ecosystems, such as wetlands and prairie remnants, that top the list of preservationists. In the future, preservation efforts will concentrate more heavily on returning damaged areas to as close to their original state as possible. This means not just setting areas aside, but reintroducing both plant and animal species that have ceased to live in a specific area, reducing pollutants and contaminants, and returning water tables to natural levels, among other things. It is a job that will become increasingly more difficult over time, as the human population continues to rise.

For those areas that already are protected, environmentalists now worry about what "protection" actually means. Areas such as parks and refuges are becoming more and more crowded with visitors. Some of the national parks are already experiencing serious overcrowding problems. Grand Canyon National Park had 2.5 million visitors in 1986, and 4.2 million by 1992, adding to smog in the canyon and creating traffic jams along the popular South Rim. Yosemite National Park had 3.8 million visitors in

1992, arriving in more than 500,000 cars and leaving 25 tons of trash and 1 million gallons of sewage each day.[40] As the wilderness continues to disappear, making sure that what is left is as clean and natural as possible will be one of the most important issues in wilderness preservation in the future.

In this century, the United States has gone from a country of vast wild expanses to a country battling over dwindling space. There are still areas to be fought over, but the next few generations may be among the last that can take an unprotected area and decide to save it unchanged for posterity. It may be well to remember the words of conservationist Stewart Udall, who said, "Posterity will honor us more for the roads and bridges we do not build . . . than those we do."[41]

Notes

1. Dyan Zaslowsky and The Wilderness Society. *These American Lands* (New York: Henry Holt, 1986), 64.

2. Ibid, 15.

3. Freeman Tilden. *The State Parks: Their Meaning in American Life* (New York: Knopf, 1962), 48.

4. Michael Williams. *Americans and Their Forests: A Historical Geography* (New York: Cambridge University Press, 1989), 3.

5. United States Department of Agriculture Forest Service. *An Analysis of the Timber Situation in the United States* (Washington, DC: GPO), 2.

6. Ibid, 3.

7. Ibid, 3.

8. Cindy D. Brown. "Mapping Old Growth." *Audubon* (May/June 1993): 132.

9. David Kelly and Gary Braasch. *Secrets of the Old Growth Forest* (Salt Lake City: Peregrine Smith, 1988), 63.

10. Lisa Couturier. "From Owls to Eternity." *E: The Environmental Magazine* (March/April 1992): 34.

11. Ibid, 34.

12. Raymond A. Young and Robert L. Geise, eds. *Introduction to Forest Science* (New York: Wiley, 1990), 431.

13. Don R. Sholly and Steven M. Newman. *Guardians of Yellowstone: An Intimate Look at the Challenges of Protecting America's Foremost Wilderness Park* (New York: Morrow, 1993), 277.

14. Margaret Fuller. *Forest Fires: An Introduction to Wildland Fire, Behavior, Management, Firefighting, and Prevention* (New York: Wiley, 1991), 98.

15. Young and Geise, 431.

16. Ross W. Simpson. *The Fires of '88: Yellowstone Park and Montana in Flames* (Helena, MT: American Geographic Publications/Montana Magazine, 1989), 21.

17. Ibid, 21.

18. Edward F. Dolan. *The American Wilderness and Its Future: Conservation Versus Use* (New York: Watts, 1992), 70.

19. Ibid, 69.

20. Jon Narr and Alex J. Narr. *This Land Is Your Land* (New York: HarperCollins, 1991), 226.

21. Jane Smiley. "So Shall We Reap." *Sierra* (March/April 1994): 81.

22. Paul Gruchow. "Rite of Spring." *Nature Conservancy* (March/April 1994): 26.

23. Ibid.

24. Jon R. Luoma. "Back Home on the Range?" *Audubon* (March/April 1993): 46.

25. Page Stegner. "Red Ledge Province." *Sierra* (March/April 1994): 98.

26. "Restoring a Thirsty Paradise." *Sierra* (March/April 1994): 125.

27. Alison Young. "Shade-Tree Mechanics." *Spirit* (February 1994): 58.

28. Ibid, 60.

29. Narr and Narr, 147.

30. National Wildlife Federation. "A Year of Crucial Decision: 25th Environmental Quality Index." *National Wildlife Magazine* (February/March 1993): 35.

31. T. E. Dahl and C. E. Johnson. *Status and Trends of Wetlands in the Conterminous United States, Mid-1970's to Mid-1980's* (Washington, DC: U.S. Department of the Interior, Fish and Wildlife Service, 1991), 1.

32. Ibid, 2.

33. Jon Kusler. "Wetlands Delineation: An Issue of Science or Politics?" *Environment Magazine* (March 1992): 9.

34. Stephen Barr. "Wetlands Excavation Rule Tightened." *The Washington Post* (26 August 1993): A12.

35. Kusler, 9.

36. Ibid, 8.

37. Stephen Barr. "Clinton To Revise Wetlands Policy." *The Washington Post* (25 August 1993): A1–A14.

38. U.S. Congress. Office of Technology Assessment. *Oil Production in the Arctic National Wildlife Refuge* (Washington, DC: GPO, 1989), 105.

39. Ibid, 106.

40. Dolan, 36.

41. Peter Wild. *Pioneer Conservationists of Eastern America* (Missoula, MT: Mountain Press, 1986), 177.

2

Chronology

EVENTS IN HISTORY THAT HAVE had the greatest impact on the American wilderness are listed here in chronological order. These include settlement, territorial acquisition, westward expansion, the creation of federal programs and agencies, the passing of legislation, and other incidents that have influenced specific policies or the philosophy concerning America's open lands. To early European settlers this philosophy began as a belief that the wilderness was an area to be conquered and tamed for economic prosperity. As the wild areas and resources were depleted, conservation and preservation became important in U.S. domestic policy and social thought.

30,000–   The first humans are believed to have arrived in North
20,000 B.C.   America from Siberia across a 100-mile land bridge over the Bering Strait. The entire continent is one vast wilderness.

1492   Columbus arrives in America. Approximately one million American Indians populate the area that encompasses the United States.

1565   The first permanent European settlement is founded by the Spanish at St. Augustine in Florida.

1607   The first permanent English settlement in North America is established at Jamestown, Virginia.

1626   The Plymouth Colony passes the first ordinances regulating the cutting of timber on colony lands.

1639   The settlement at Newport, Rhode Island, prohibits deer hunting for six months of the year.

1700   About 250,000 people live in the original 13 colonies, from Georgia to Maine.

1770   Due to the influx of settlers and the high birthrate, the population of the 13 colonies numbers approximately 2 million people.

1778   After the Revolutionary War with England is over, many of the newly formed United States claim territories to their west, outside of actual state boundaries. The state of Maryland, holding no such claim, complains that all state-owned western territories should be turned over to the federal government.

1780   The Continental Congress agrees with Maryland and requires all states to turn over annexed territories to the federal government, creating the first federally owned public lands.

1803   President Thomas Jefferson acquires the Louisiana Territory from France for $27 million, doubling the size of the United States.

1805   The Lewis and Clark expedition reaches the Pacific Ocean in Oregon.

1812   The General Land Office is established as part of the Treasury Department in order to handle all sales of federal lands.

1826   Jedediah Strong Smith crosses the southwestern deserts and becomes the first American to reach California over land.

1841   The Preemption Act, the first U.S. land law in which the federal government sets guidelines for the sale of public domain lands, is passed. Anyone, as long as they either are a U.S. citizen or intend to become one, is allowed to go into the unsurveyed lands of the West and stake a claim for $1.25 per acre. The price is extremely low because the government is promoting western settlement.

1846 The United States and Britain sign the Oregon Treaty, giving the United States full claim to the Oregon Territory, including Oregon, Washington, Idaho, and parts of Wyoming and Montana.

1848 The United States wins the Mexican War and takes control of the California and New Mexico territories, containing what are now the states of California, Nevada, Utah, and parts of Arizona, New Mexico, and Colorado.

1849 The Department of the Interior is established as a cabinet level office in order to centralize several disparate agencies that include the General Land Office, the Bureau of Indian Affairs, and the Patent Office.

1862 The Homestead Act opens public domain lands to unrestricted settlement. Any citizen over 21 years of age, or the head of a household, can claim up to 160 acres of federal land, live on their claim for six months, and then buy it for $1.25 an acre; or cultivate the land and live on it for five years, after which they can own it for the price of a small filing fee.

1867 The United States purchases Alaska from Russia for $7.2 million—about 2 cents per acre.

1872 The General Mining Law opens all public domain lands to private prospecting and development. By discovering minerals on a piece of land, paying for a boundary survey, and applying to a land office, a person can purchase the land for $2.50 per acre for surface mining and $5.00 per acre for underground mining. All that is required is for the owner to invest $500 into the land within five years of staking the claim.

Congress establishes Yellowstone as the first national park in the world.

1877 The Desert Land Act authorizes the purchase of up to 640 acres of public domain land by settlers in the desert

1877
cont.

Southwest for $1.25 an acre, if the settler agrees to irrigate the tract within three years.

1890

Yosemite, Sequoia, and General Grant (now part of Sequoia) are established as permanent national parks, the first since Yellowstone 18 years earlier.

1891

The General Revision Act repeals the Preemption Act, reduces the number of acres allowed in the Desert Land Act from 640 to 320, and puts limits on the auction sale of public land. Perhaps most significantly, the act allows the president to withdraw from settlement any area that the Secretary of the Interior determines is in need of watershed protection and timber preservation. This act signifies a change in the way the federal government perceives the public lands—from an inexhaustible to a limited resource.

The Forest Reserve Act authorizes the president to "set apart and reserve . . . any part of the public lands wholly or in part covered with timber or undergrowth whether of commercial value or not, as public reservations." President Benjamin Harrison proclaims 13 million acres in Wyoming and Colorado as reserves. President Grover Cleveland later adds 20 million more acres between 1893 and 1897, even though the purpose of the reserves is not clear.

1892

The Sierra Club, the first major private organization in the world dedicated to preserving wilderness, is founded by John Muir to help protect the Sierra Nevada Mountains from resource development. Muir becomes the organization's first president.

1897

The Forest Organic Act is passed in order to ensure a continuous supply of water and timber in the United States. This act establishes forest reserves for the purpose of resource management, placing reserves under the control of the General Land Office of the Department of the Interior.

1902

The Newlands Reclamation Act authorizes financing and construction of federal irrigation projects on public lands.

1905   Under the Reorganization Act, the United States Forest Service (USFS) is established as part of the USDA. All forest lands that had been declared as forest reserves by presidential decree under the Forest Reserve Act are to be managed by the USFS.

1906   Largely as a result of thefts of Anasazi Indian relics in the Southwest, Congress passes the Antiquities Act, authorizing the creation of national monuments. These sites are areas of scientific, historic, or scenic value. The president is authorized to designate a monument by executive order, allowing immediate protection of a site and avoiding political debate and possible rejection by Congress.

President Theodore Roosevelt withdraws 66 million acres of coal lands on which hundreds of claims are pending in order to prevent what is becoming a monopoly by a few small coal companies. According to law, 160-acre claims can be filed for $10 per acre, but no more than four claims can be owned by a single company or individual. A few companies have entered scores of claims under false names, though, and own most of the 30 million acres that have been parceled out. President Roosevelt's decision meets with outcry from many, but sets a precedent for the president to set certain public domain lands off-limit from land law policies.

1910   The Pickett Act allows the president to withdraw any portion of the public domain from homesteaders for "any public purpose." This law is meant to give the government more control over the land grab that persisted into the twentieth century. The act followed on the heels of and was influenced by the 1906 controversy over coal lands.

1913   The city of San Francisco had wanted to build a dam over the Tuolumne River in Yosemite National Park since the late 1800s in order to provide hydroelectric power, but such a dam would also flood the Hetch Hetchy Valley, which John Muir considers to be one of the most spectacular areas in the park. Congress agrees to the proposal anyway, but two succeeding secretaries of the Interior Department refuse to allow the dam to be built. Finally, in 1913, former San Francisco city attorney

1913 *cont.*	Franklin Lane is appointed Secretary of the Interior, and gives the plan the green light. John Muir considers this his greatest defeat and is frequently ill after the dam's approval; he dies in 1914. Within a few years the Hetch Hetchy Valley is under water.
1916	The National Park Organic Act establishes the National Park Service (NPS) as part of the Department of the Interior and sets guidelines for managing national parks. Before this act passed there were already 15 national parks with no agency to administer them.
1920	The Mineral Leasing Act sets up a system in which oil and mineral reserves on the public domain, national forests, and wildlife refuges can be leased to private companies on a competitive bidding system. If the government decides to open an area for extraction of resources, private firms can bid for the projects, with both the state and federal governments receiving a cut in royalty payments.
1924	Persuaded by Aldo Leopold, the USFS sets aside 500,000 acres at the headwaters of the Gila River, New Mexico, and creates the first officially designated wilderness area in the United States.
1934	The era of the indiscriminate settlement by homesteaders finally comes to a close with the passage of the Taylor Grazing Act. This ends land claim settlements on public domain land in nine western states and Alaska. It also places 142 million acres of federal land into grazing districts, on which ranchers can lease areas for grazing their cattle, goats, or sheep. The original fee is five cents per cow, per month. Though this fee is extremely low, even for the time, ranchers still complain about it bitterly.
1939	The Reorganization Act takes the Bureau of Fisheries from the Department of Commerce, and the Bureau of Biological Survey from the Department of Agriculture (USDA), and combines them to form the United States Fish and Wildlife Service (USFWS) under the Department of the Interior. Wildlife refuge administration is one of the new agency's main responsibilities.
1946	The Department of the Interior's General Land Office and Grazing Service are combined to form the Bureau

of Land Management (BLM). The new agency will administer all of the areas now known as National Resource Lands.

1950   Newton Drury, director of the NPS, is looking through a copy of the Federal Register when he notices that the Bureau of Reclamation is planning to build two dams in Dinosaur National Monument in Colorado and Utah, even though he has not been consulted and knows nothing of the project. Drury is infuriated, but Secretary of the Interior Oscar Chapman supports the project. Drury's stalwart objections to the idea only lead to his forced resignation, but a public outcry is raised and a coalition of conservation groups join together to fight the planned dams. These groups include the Izaak Walton League, the Sierra Club, and The Wilderness Society. This marks a new level of cooperation among environmentalists.

1956   Due to the public pressure, and after heated debate, Congress finally votes to kill plans to build dams in Dinosaur National Monument.

1963   Congress passes the first Clean Air Act to send funds to individual states for use in reducing air pollution. The bill is amended in 1970 to set specific standards for lowering lead, carbon monoxide, and hydrocarbon emissions.

1964   The National Wilderness Preservation Act is passed by Congress to establish a system of wilderness areas in which each addition to the system is completely protected from any form of permanent human development. Known more commonly as simply the Wilderness Act, it is signed by President Lyndon B. Johnson and is a landmark in conservation legislation. The act provides for the protection of 9 million acres of wilderness and calls for the NPS, the USFS, and the USFWS to review their jurisdictions for additional wilderness areas (see Chapter 1, p. 12).

1964–1966   The Bureau of Reclamation announces plans to build two major dams on the Colorado River that will flood large portions of the Grand Canyon, including areas within Grand Canyon National Park. The idea is met

1964–1966
cont.

with extreme public criticism. The same coalition of organizations that had cooperated in fighting the Dinosaur National Monument dams in 1956 rally together against this new plan, which is eventually dropped because of the outcry.

1966

The National Wildlife Refuge System Administration Act consolidates all federal wildlife refuges into the National Wildlife Refuge System, managed by the USFWS. The act does not provide special protection for the refuges, but is primarily designed to make them easier to administer.

1969

The National Environmental Policy Act (NEPA) requires all federal agencies to consider the environmental effects of their policies through certain procedures, such as conducting Environmental Impact Studies for any program that might affect the environment.

1972

The Federal Water Pollution Control Act, known as the Clean Water Act, sets guidelines for reducing pollution into the nation's navigable waters. It also requires that before anyone can alter a wetland, they must first obtain a permit from the Army Corps of Engineers. Some developers find a loophole to this section of the law, forcing President Clinton to impose stricter restrictions in 1993.

The USFS initiates the Roadless Area Review and Evaluation (RARE) to survey their roadless lands for possible additions to the National Wilderness Preservation System. Fifty-nine million acres are surveyed, and 12.3 million, or 19 percent, are recommended for preservation.

1973

Congress enacts the Endangered Species Act to help protect threatened species from extinction due to human development. The act makes it illegal to harm any species placed on the "endangered" or "most threatened" list. The law gives environmentalists a powerful tool to use, not only to save species, but also in defense of entire ecosystems. In southern California during the early 1990s, for example, local environmentalists oppose a housing and road development on shrinking coastal open space. When they find that it threatens the California gnatcatcher, a species on the threatened list, the project has to be scaled back considerably.

1974   The Forest and Rangeland Renewal Act sets guidelines for long-range and continuous inventory of all federal, state, local, and privately owned rangeland and forest resources.

The Eastern Wilderness Act sets aside 16 areas, totaling 207,000 acres for preservation under the National Wilderness Preservation System. Many of these areas were previously damaged by man but were in advanced stages of recovery. The act sets a precedent by showing that an area does not have to be "untouched" by man to be given wilderness protection.

1976   The National Forest Management Act requires the USFS to create 50-year management plans for all of their lands. These plans include economic, recreation, wilderness, and wildlife studies.

The Federal Land Policy and Management Act creates public land policy guidelines for all BLM lands. This is the first act that clearly defines the mission and authority of the BLM. The act also repeals all existing public land laws except the General Mining Law of 1872, and calls for the BLM to review all of their holdings for possible additions to the National Wilderness Preservation System.

1977   After criticism that the Forest Service RARE did not go far enough, the service initiates a second survey, RARE II, which identifies 62 million acres of potential wilderness areas.

1978   The National Parks and Recreation Act of 1978 sets guidelines for creating more national parks, wilderness areas, trails, and wild and scenic rivers. It creates such parks as Santa Monica Mountains National Recreation Area (California), Statue of Liberty National Monument (New York), and Guam National Seashore.

1980   The Alaska National Interest Lands Conservation Act (ANILCA) creates a plan for conservation of public lands in Alaska by placing public land into national parks

1980
cont.

and preserves, national forests, National Wild and Scenic Rivers, national wildlife refuges, and the National Wilderness Preservation System. The act adds 104.3 million acres to these conservation systems and designates 56 million of those acres as wilderness areas.

1987

Three workers at a sawmill in Cloverdale, California, are killed when their saw shatters after hitting a spike driven into a tree by an unknown radical environmentalist. "Tree spiking" is a method used by some militant environmentalists to keep loggers from cutting forests. Spikes are driven into trees to make them unsafe to cut down, though usually the spiked trees are marked with paint to make loggers aware of the danger. This incident sparks outrage from loggers and mainstream environmental groups alike, who argued that these destructive and dangerous tactics hurt both sides of the conflict. After this accident, the radical environmental group, Earth First!, stops sanctioning tree spiking as a viable tactic and spiking declines dramatically, though still occurring periodically.

1990

Congress passes a second Clean Air Act. Proposed by President Bush the year before, the new act sets guidelines for an overall reduction in sulphur emissions in the United States by ten million tons per year by the year 2000, and nitrogen oxide emissions to be cut in half. It calls for a reduction in automobile emissions, and requires the 111 largest coal burning electric power plants in 22 states to cut their pollution levels by such means as switching to coal with a lower sulphur content or using sulphur cleansing technologies in their smokestacks.

1991

U.S. District Judge William L. Dwyer halts logging across millions of acres of national forests in the Pacific Northwest until the USFS can devise a plan to save the northern spotted owl. The USFS never completes such a plan. The Clinton Administration produces their own plan in 1993.

1993

In April, at a Forest Summit in Portland, Oregon, President Bill Clinton and Vice President Al Gore meet with environmental and logging industry leaders during a one day conference to discuss the future of logging in the Pacific Northwest. No specific decisions are made, but this meeting marks a point of optimism as the two opposing sides came together to discuss solutions to their conflict. President Clinton orders his cabinet members to draft a plan to be presented within 60 days.

President Clinton's promised plan to save the northern spotted owl and the timber industry in the Pacific Northwest is presented in July. The plan calls for a reduction in the timber harvest to 1.2 billion board feet per year, down from 5 billion in the mid-1980s, and includes a $1.2 billion aid package to help the economy recover from a loss of timber jobs (see Chapter 1, p. 17).

Also in July, the Clinton administration and the sugar industry in Florida agree on a $465 million plan to save the Florida Everglades. Secretary of the Interior Bruce Babbitt calls the plan, "the largest, most ambitious ecosystem restoration ever undertaken."

In August the Clinton administration announces a new wetlands policy with a goal of "no overall net loss of wetlands," reversing a proposal by President Bush that would have allowed for development on 1.7 million acres of land in Alaska. The USDA Soil Conservation Service is put in charge of administering the new policy. As with all plans in the past, the effectiveness of this one depends on the definition of a wetland, which the National Academy of Sciences is to produce in 1994. The new policy also includes regulations to close a loophole in the Clean Water Act of 1972 (see Chapter 1, p. 29–31).

In October the National Biological Survey brings together approximately 950 biologists for a historic attempt to inventory all of the plants and animals in the nation, with the understanding that before we can preserve our wildlife and habitats we must be fully aware of what exists.

In November Congress attempts to overturn the last federal land law still in effect, the General Mining Law of 1872. Both houses of Congress pass similar measures to rescind the General Mining Law but cannot agree on a common version of the bill by the end of the session. Election of a Republican majority in 1994 puts the idea on hold.

1994   The U.S. Congress passes the California Desert Protection Act. The law designates nearly 4 million acres of BLM lands as wilderness areas. It also upgrades Joshua Tree and Death Valley from national monuments to national parks, creates the new 1,458,000 acre Mojave National Park, and designates nearly 4 million acres of park service and 9,000 acres of USFS lands as wilderness areas.

In an effort to save declining salmon stocks, the Pacific Fishery Management Council places a ban on salmon fishing off the coast of Washington State and initiates tight restrictions on salmon fishing off the coasts of Oregon and California.

3

Biographical Sketches

THIS SECTION IS A SELECTED LIST of some of the most influential people in the field of American wilderness conservation and preservation, both current and historical. For more comprehensive information, the reader is advised to consult the biographies listed in chapter six.

Edward Abbey (1927–1989)

Edward Abbey was born and raised on a farm just outside Home, Pennsylvania. In 1944, at the age of 17, he went on a hitchhiking trip to the American West where he saw the striking beauty of the region's mountains and deserts for the first time. The following year Abbey was drafted by the U.S. Army and sent to Italy for a two-year tour of duty. In 1947 he returned to the United States and began studying at the University of New Mexico, where he eventually received a bachelor's and a master's degree in philosophy.

After graduating, Abbey began a writing career that would eventually make him one of the best known environmental writers in the nation. His first novel, *Jonathan Troy,* was published in 1954, and over the next ten years he went on to publish a string of novels, including *The Brave Cowboy* (1956) and *Fire on the Mountain* (1962). Unable to make a living at writing, however, Abbey moved west and became a ranger at Organ Pipe Cactus National Monument in southern Arizona. He went on to work as a fire lookout and ranger at the Grand Canyon, Arches National Monument in Utah, and Glacier National Park in Montana.

In 1968 Abbey achieved his first broad success as an author with the publication of *Desert Solitaire: A Season in the Wilderness,* a collection of ruminations about his time spent at Arches National Monument, his philosophy on wilderness, and most notably, angry attacks on society's ignorance and abuse of nature. Suddenly, Abbey was famous and magazine and book publishers eagerly solicited his work. He wrote the text for several large coffee table books with wilderness photographs, including *Shiprock: The Canyon Country of Southeast Utah* (1971), *Cactus Country* (1973), and *Appalachian Wilderness* (1973).

The novel which most exemplifies Abbey's philosophy and persona is *The Monkey Wrench Gang* (1975). This story follows a fictional band of rag-tag eco-terrorists as they travel the West blowing up bridges and destroying development projects in the name of the environment. To a large extent this book solidified a whole new way of thinking that would later be taken to heart by a generation of environmentalists frustrated at fighting what seemed like a losing battle. To people like David Foreman and members of his Earth First! movement, the time for simple negotiation alone was passed, and to really achieve results one had to take the environmental fight to a physical level. The writings of Edward Abbey both inspired and gave voice to those frustrations, and to radical environmentalists he became a hero for publicly expressing what many felt.

Ansel Adams (1902–1984)

Born and raised in San Francisco, California, Ansel Adams received only minimal formal schooling, relying on private instruction in his home and self-education. In his youth Adams visited Yosemite National Park every summer. Later he spent four years as the head of the Sierra Club's Yosemite headquarters. His first portfolio of original photographs, "Parmelian Prints of the High Sierra," was issued in 1927, launching his career as a wilderness photographer. In 1941 he was appointed to the position of Photomuralist for the Department of the Interior.

In 1956–1957 Adams was president of an environmental organization, the Trustees for Conservation. He continued to work for the Sierra Club for his entire life, and in 1963 he received the John Muir award from the club for his services to conservation. Adams went on to become one of the best-known wilderness photographers of all time. His pictorial books include *Illustrated*

Guide to Yosemite Valley (1930), *My Camera in the National Parks* (1949), and *The Parklands of America* (1962).

John James Audubon (1785–1851)

Born in Santo Domingo (now known as Haiti), John Audubon was raised in France. He first came to the United States with his family in 1803 and took up an interest in studying, drawing, and painting birds. In 1824 he took his portfolio of watercolor paintings to the Philadelphia Academy of Sciences and was advised to publish them in England. The engraving equipment was better than in America, and in England he would not have to compete with another influential ornithologist, George Ord. Audubon's paintings were published in four volumes under the title *The Birds of America*, from 1827 to 1838. Accompanying written text was published separately in five volumes under the title *Ornithological Biography*, from 1831 to 1839. Audubon went on to become the most famous ornithologist in the world. He died before the conservation movement developed, but one of the largest environmental organizations in the world, the National Audubon Society, was founded in his name in 1905.

Bruce Edward Babbitt (1938–   )

Born into a well-to-do family in Flagstaff, Arizona, Bruce Babbitt's family made its fortune raising cattle and operating trading posts. Babbitt majored in geology at the University of Notre Dame, graduating magna cum laude in 1960. He went on for a master's degree at the University of Newcastle in England and planned a career in the mining industry, but changed his mind and went to Harvard Law School instead. Upon graduation he returned to Arizona and became a lawyer. In 1974 Babbitt was elected Arizona's attorney general, where he battled organized crime, public corruption, and land fraud. Four years later he was elected governor; a position he held until 1988, when he lost a re-election bid.

In 1979 Babbitt was appointed to the President's Commission on the Accident at Three Mile Island. In 1980 he was on the Nuclear Safety Oversight Committee. In 1981 he received the Thomas Jefferson Award from the National Wildlife Federation for his work in conservation. From 1990 to 1992 Babbitt was president of the League of Conservation Voters. He was often critical of the Department of the Interior during both the Reagan and Bush administrations.

When Bill Clinton was sworn in as president in 1993, Babbitt was appointed Secretary of the Interior. His appointment was met with cheers from the environmental community as his views were much more environmentally friendly than his predecessor, Manuel Lujan. Also, with a background in ranching and a good degree of popularity in the West, he was seen as a man who could bring environmental and business groups together.

Babbitt supported using the Endangered Species Act to protect entire ecosystems, not just focusing on specific species. He also supported a reduction in the timber harvest, reintroducing wolves to Yellowstone National Park, limiting development in the national parks, and blocking oil drilling in the Arctic National Wildlife Refuge. In his first year with the Interior Department, Babbitt initiated the largest ecosystem recovery plan in history to protect Florida's Everglades.

David Brower (1912–   )

David Brower was born and raised in Berkeley, California, and spent many of his early days wandering in the Berkeley hills and hiking with his family in the nearby Sierra Nevada mountains. He dropped out of the University of California at Berkeley during his sophomore year and spent as much time in the Sierras as he could, working first at a camp near Lake Tahoe and later as an employee of Yosemite National Park.

In 1933 Brower joined the Sierra Club and began leading trips into the mountains. He became increasingly involved in the club's activities, rallying for conservation causes and joining the club's editorial board. When World War II broke out, however, Brower's Sierra Club activities were put on hold as he volunteered for the army's Tenth Mountain Division and was sent overseas to fight in Italy.

When World War II ended, the 60-year-old Sierra Club was experiencing a difficult stage. The club was composed entirely of volunteers, but the conservation battles they were beginning to take on severely taxed the volunteers' time and energy. For this reason, David Brower was named the Sierra Club's first paid executive director in 1952.

Brower used the club's influence to fight many conservation battles, including his most important: to stop plans by the Bureau of Reclamation to dam the Colorado River and flood parts of the Grand Canyon. Without authorization from the Sierra Club's board of directors, Brower took out a costly full-page ad against the idea in the *New York Times,* which asked in part, "Should we also flood the Sistine Chapel so tourists can get nearer the ceiling?" The ad successfully united public opinion against the project, but his tactics caused friction with the Sierra Club and the board of directors.

Under Brower's leadership the Sierra Club fell deeply into debt, with a deficit of $129,000 in the first ten months of 1968. Due largely to the *New York Times* ad they also lost their tax-exempt status. Some argued that this was a blessing in disguise, because now the club could operate with more lobbying power and without restrictions on their activities, but the board of directors had financial concerns. Though he had made great gains for wilderness preservation during his short tenure as Executive Director, the board decided that the organization could not continue to operate under his leadership and he was asked to resign. Brower remained active in conservation and went on to found three of his own conservation organizations: The John Muir Institute in 1968, Friends of the Earth in 1969, and Earth Island Institute in 1981.

John Burroughs (1837–1921)

Born in the Catskill Mountains near Roxbury, New York, John Burroughs was a wilderness writer who grew to be a legend. In his late teens, Burroughs attended the Hedding Literary Institute and Cooperstown Seminary for six months before becoming a teacher. In 1863 he moved to Washington, D.C., secured a job with the Treasury Department, and started spending time with the poet Walt Whitman, one of his greatest mentors. In the meantime, Burroughs was becoming quite well known as a nature writer, with articles in numerous magazines.

In 1871 Burroughs published his first book on nature, *Wake Robin.* He went on to write a total of 27 books, including *Winter Sunshine* (1875), *Locusts and Wild Honey* (1879), and *The Light of Day* (1900). Burroughs was quite a character himself, whose personality bolstered his fame. In his later years he had long gray hair and an equally long, bushy gray beard and moustache. For a time he was the most photographed man in the nation, and he frolicked across the country on often zany camping trips with the likes of Thomas Edison, Harvey Firestone, Henry Ford, John Muir, and Theodore Roosevelt. Burroughs inspired generations with a love of nature and helped create a conservation mentality in some of the most influential minds of his day.

Marjory Stoneman Douglas (1890–   )

Marjory Stoneman Douglas was born in Minneapolis, Minnesota, but her parents divorced when she was six and Douglas moved with her mother to live with her grandparents in Massachusetts. It was here that she was raised, later attending Wellesley College just outside of Boston. A few weeks after she graduated, in 1912, Douglas's mother, Lillian Stoneman, died of cancer. During the ensuing period of grief and loneliness, Marjory Stoneman was briefly married to Kenneth Douglas before divorcing him and moving in with her father, Frank Stoneman, who had since moved to Miami and founded the *Miami Herald* newspaper. Marjory worked as a reporter for the paper, and began taking trips to watch the sunrise over the Everglades. It was these early morning excursions that would inspire her to dedicate her life to preserving this unique region.

Douglas went to France during World War I to work for the Red Cross, returning to Miami to write fiction and teach in 1920. In 1947 she published her most famous work, *The Everglades: River of Grass,* and began to concentrate her effort to save this wilderness from developers who wanted to drain it. In 1970 she founded and became president of Friends of the Everglades. She also led the Coalition to Repair the Everglades. In 1975 she was named Conservationist of the Year by the Florida Audubon Society, and a year later she was given the same title by the Florida Wildlife Federation. More than anyone else, it was Marjory Stoneman Douglas who demonstrated the unique natural value of the Everglades and inspired the campaign to save it. Douglas continued to speak at public forums any time the Everglades were threatened well into her 90s, gave interviews into her 100s, and became known warmly as the "Grand Damme of the Glades." Douglas was awarded the Medal of Freedom by President Bill Clinton in December of 1993.

William O. Douglas (1898–1980)

At the age of three William O. Douglas was afflicted with polio. His leg muscles withered away and he was completely unable to walk. Fortunately, he recovered from the disease and learned to walk again, but when Douglas was six his father died and the family was forced to live in poverty, scavenging for junk in alleys and garbage cans to survive. In order to escape this dreary life, Douglas went on excursions into the mountains near his hometown of Yakima, Washington. Instead of hiking through the mountains, however, Douglas ran, often becoming nauseous from the exertion, yet he still managed to find a sense of solace in nature.

After high school, Douglas accepted a scholarship at Whitman College in Walla Walla, where he rode his bike 165 miles each way to school and back every day, and graduated in 1920. In 1925 he graduated from Columbia Law School and went to work for a New York law firm. In 1936 he was appointed to the Securities and Exchange Commission by President Franklin Delano Roosevelt, and three years later, in 1939, to the Supreme Court. It was in this post that Douglas made his greatest contributions to conservation in America.

William O. Douglas wrote several books on the wilderness, including *A Wilderness Bill of Rights* (1965) and *Farewell to Texas: A Vanishing Wilderness* (1967). He was a conservation activist, marching for environmental causes and making public speeches against USFWS policies in the Wind River Mountains and an NPS plan to create towns in Yellowstone. Douglas made many enemies among pro-business conservatives and several unsuccessful attempts were made to impeach him from his Supreme Court post, including one in 1970 led by Gerald Ford and Richard Nixon.

In 1972 Douglas created a legal precedent that gave conservationists a new tool in the battle to save the wilderness. In the case of *Sierra Club v. Morton* he wrote that people who have a meaningful relationship to a natural entity should have the right to speak for and represent that entity's right to exist in a court of law. The Sierra Club sued as a representative of Mineral King Valley, on which the Walt Disney Company wanted to build a 20-lift ski resort. Douglas wrote that since courts regularly treat ships, corporations, and other inanimate objects as personalities and grant them legal rights, objects in nature should have similar legal rights. The Sierra Club lost its case 4–3, but through Douglas's dissent a new legal way of thinking was born. Douglas retired from the bench in 1975.

David Foreman (1946–   )

David Foreman was born and raised in New Mexico. After graduating from the University of New Mexico in 1968, he went to the

Marine Corps Officer Training School in Quantico, Virginia, but only lasted 61 days, 31 of which were spent in the brig in solitary confinement before he was discharged. In the mid-1970s he worked as the southeast regional representative for The Wilderness Society. Later he moved to Washington, D.C., to become the organization's chief lobbyist.

In 1980 Foreman left The Wilderness Society to form his own environmental movement, Earth First! Foreman and his movement were the source of considerable controversy even among environmentalists due to his promotion of monkeywrenching as a viable tool in fighting development. Monkeywrenching involves such techniques as pouring foreign substances in the gas tanks of heavy machinery, drilling spikes into trees so that they cannot be cut safely, and other procedures designed to attack developers directly. This type of action was taking the environmental fight to a whole new level that most environmentalists did not agree with. In 1985 Foreman cemented his role as the leader of the movement with the publication of his how-to book, *Ecodefense: A Field Guide to Monkeywrenching.*

Foreman left the Earth First! movement in 1990 amid continuing negative publicity. Since that time he has endorsed numerous environmental causes throughout the United States, rallying support through public speaking and other legal means. He is also the author of several other books including a descriptive inventory of the larger wilderness areas, *The Big Outside,* and *Confessions of an Eco-Warrior* (1991).

George Bird Grinnell (1849–1937)

The son of a wealthy stockbroker, George Bird Grinnell was born in New York City. In 1857 the family moved to Audubon Park, the estate of the late John James Audubon, in what was then a wooded section of upper Manhattan. Grinnell grew to be quite close to the surviving Audubon family, playing with John James's grandson and attending a school conducted by Audubon's widow.

When George Bird Grinnell went off to Yale to study paleontology, he was one in a group of 12 students to accompany Professor Othniel Marsh on a much celebrated expedition to the West to study fossils in 1870. The group traveled to Fort McPherson, Nebraska, where they met Buffalo Bill Cody, were protected from Indian attack by armed soldiers, and collected data to support Charles Darwin's theory of evolution. Hooked by the excitement of his adventure, Grinnell continued to return to the West nearly every summer for the next 40 years to hunt and gather scientific evidence. He was invited to join General George Custer's infamous last expedition, but had to decline the offer when he went to work for the Peabody Museum.

As Grinnell matured he became more and more concerned with conservation in America. In 1875 he joined a column of soldiers on a survey expedition to Yellowstone National Park, which had yet to be fully explored. As the trip's naturalist, Grinnell was astonished to find the park full of poachers who were killing thousands of buffalo each year and leaving their carcasses behind to rot. Grinnell reported to Congress that unless something was done, it would not take long before the once abundant buffalo herds became extinct.

In 1880, at the age of 31, Grinnell took over as the editor of the magazine *Forest and Stream*. For the rest of his career, Grinnell would use this post as a platform to effectively voice his environmental concerns. Using the vanishing buffalo to rally the sentiment and support of the public, Grinnell launched a crusade for legislation to protect all of Yellowstone's animals. The result was the Act to Protect the Birds and Animals of Yellowstone Park, passed by Congress in 1894. Grinnell also became a close friend and advisor to New York's Governor Theodore Roosevelt, and is credited for a great influence on the future president's role in conservation. In 1905 Grinnell founded the first Audubon Society, an organization dedicated to wilderness preservation.

Harold Ickes (1874–1952)

Born in Pennsylvania, Harold Ickes moved to Chicago after his mother died when he was 16. Ickes finished high school and attended the University of Chicago, graduating at the top of his class in 1897. He worked as a journalist before going on to receive a law degree, also from the University of Chicago. Ickes was active in politics and worked on Theodore Roosevelt's presidential campaign in 1912. He went on to become a leader in Roosevelt's Progressive political party and a champion of the conservation movement. In 1932 he led the National Progressive League in support of Franklin Delano Roosevelt's presidential bid. When Roosevelt won his campaign, in a surprising move to all, including Ickes, he appointed Ickes to head the Department of the Interior.

Ickes held the post for more than 12 years—longer than anyone before him or since.

Under Ickes's control, more than 18 million acres were added to the national forests, including several new areas to the national parks. With President Roosevelt's New Deal, two enormous projects were born: the Civilian Conservation Corps (CCC) and the Tennessee Valley Authority (TVA). The CCC put millions of men to work across the country on soil conservation projects. The TVA built dams throughout the Tennessee River Valley and became one of the administration's most controversial programs. Ickes retired from office in 1946 when President Harry S Truman succeeded Roosevelt.

Aldo Leopold (1887–1948)

Aldo Leopold was born in Burlington, Iowa. As a child he explored the marshes and forests near his home, became an avid hunter, and decided early in life that he wanted a career in forestry. In 1909 he graduated with a master's degree from the Yale University School of Forestry and went to work for the USFS in the Arizona and New Mexico territories.

Leopold is perhaps best known for his journals, which have inspired generations of conservationists. His collection of essays, *A Sand County Almanac,* has sold more than one million copies since its publication in 1949—it shows a man whose views on conservation and wilderness management evolved over the years.

The USFS, for which Leopold worked, has always been concerned primarily with the extraction of resources, namely lumber. In his day the idea to set aside national forest land for preservation was unprecedented. In 1924, however, his persistence led to the creation by the USFS of the first official wilderness area in the United States, the 500,000-acre Gila River Wilderness Area in New Mexico. It was Leopold's desire to keep out all roads and human construction in order to create a "national hunting ground."

Part of his plan would later backfire, when, in 1928, he helped lead a crusade to exterminate wolves and mountain lions from the area in order to increase the number of deer. He hoped to create a "hunting paradise," but the end result was the proliferation of deer to dangerously high numbers. A road was cleared into the area to allow hunters better access in order to thin the deer population. The failure of his policy had a great effect on Leopold, reshaping the way he thought about wilderness management. He was as saddened by the loss of the wolves and mountain lions as by the mismanagement of the deer. Leopold's view of wilderness evolved from merely a place to hunt to a place to preserve in its natural state.

For the rest of Leopold's career he advocated a change in attitude toward the wilderness by the public. He worked, through his writings and teaching, to educate the public to respect the land while there was still wilderness left to respect. In 1933 he became the first professor of game management in the United States at the University of Wisconsin.

Manuel Lujan Jr. (1928–   )

After working in the insurance business, Manuel Lujan began a career in national politics when he was elected to Congress in 1969. The New Mexico Republican served for 20 years. As the ranking Republican on the House Interior Committee in 1982, Lujan made headlines when he angrily introduced a resolution to stop then Secretary of the Interior James Watt from issuing oil and gas leases in wilderness areas in the western states. In 1988 Lujan was appointed Secretary of the Interior by President George Bush. Lujan only accepted the post reluctantly after a personal appeal from President Bush.

Despite his reaction to the western oil and gas leases, Lujan's policies during his term as Interior Secretary were predominantly pro-business. His most controversial decision came in May 1992 when he was instrumental in granting an exemption from the Endangered Species Act on timber sales from various federal lands in Oregon, within the habitat of the endangered northern spotted owl. This was only the second exemption ever granted to the act since it was created in 1973. Lujan has long opposed the Endangered Species Act, citing detrimental effects to American business.

Lujan also supports oil drilling in the Gulf of Mexico and in the Arctic National Wildlife Refuge in Alaska. On the environmental side, Lujan supports a ban on offshore drilling off most of the California coast. He has also worked to improve schools for American Indians, to protect historic battlefields, and to improve the Department of the Interior's record on minority hiring.

George Perkins Marsh (1801–1882)

As a child in Woodstock, Vermont, a temporary visual disability forced George Perkins Marsh to give up a love of reading and turn his attention to the forests and wilderness surrounding his home. In 1816, at the age of 15, Marsh entered Dartmouth College. After graduating, he taught Greek and Latin, tried a few unsuccessful business ventures, became an attorney, and went into politics. In 1840 he was elected to Congress. where he was largely responsible for the creation of the Smithsonian Institute.

In 1857 he wrote a report tying the deterioration of Vermont's farmlands and forests to the decline of fish in the state's rivers, streams, and lakes. This report was revolutionary for its time, as it showed that man's impact on one aspect of nature was inextricably tied to an entire system. In 1863, while serving as minister to Italy, Marsh completed his crowning achievement: the book *Man and Nature; or, Physical Geography as Modified by Human Action.* This highly regarded text laid a foundation for the entire conservation movement by calling into question man's impact on the world around him. Marsh was one of the first to demonstrate mankind's destructive influence on the rest of the natural world. Marsh proved to be ahead of his time, however, as his warnings went largely unheeded until the start of Theodore Roosevelt's conservation movement 40 years later.

Robert Marshall (1901–1939)

Born and raised in New York City, Robert Marshall went on to obtain a bachelor of science degree from the New York State College of Forestry in 1924, a master of science from Harvard Forest School in 1925, and a doctorate in plant physiology from Johns Hopkins University in 1930.

Marshall was one of the pioneer conservationists in the United States. He was an avid hiker who loved the outdoors and devoted his life to preserving it through lobbying and government legislation. He worked as a researcher for the USFS in the Rocky Mountains, Montana, and arctic Alaska. In 1933 he was appointed director of the Office of Indian Affairs, a division of the U.S. Department of the Interior, where he worked to help expand the American Indians' role in managing forests on their reservations. In 1937 he returned to the USFS as chief of the Division of Recreation and Lands.

It was Marshall who was responsible for the first set of regulations for creating wilderness areas. Under Marshall's plan, roughly 14 million acres of land were originally labeled as *primitive areas.* Any areas determined to be over 100,000 acres after further study were designated *wilderness areas*. These regulations eventually evolved, with considerable help from Howard Zahniser, into the Wilderness Act, passed by Congress in 1964.

Marshall was the author of many books on wilderness conservation, including *The Social Management of American Forests* (1930), *The People's Forests* (1933), and *North to Doonerak* (1939). In 1935 he helped found The Wilderness Society.

Stephen Mather (1867–1930)

The son of a well-to-do businessman, Stephen Mather was born and raised in California. After graduating from the University of California at Berkeley, Mather went on to create a name for himself as a reporter on the *New York Sun* before joining his father's borax business.

In 1914, Mather took a trip to Yosemite Valley and was amazed at what he considered pathetic park management, with cattle grazing throughout the park and some of the most scenic areas sold off to private individuals. To voice his concern, Mather immediately wrote a letter to Secretary of the Interior Franklin K. Lane. Lane was also a Californian, and a graduate of the University of California at Berkeley, and he immediately recognized Mather as his schoolmate. He returned Mather's letter just as quickly, inviting Mather to come to Washington to run the park system. Mather agreed, moved to Washington as Lane's assistant, and began what would become a lifelong career in wilderness management and conservation.

When Mather went to work for the Department of the Interior, his first goal was to increase the amount of funding Congress appropriated to the National Park System. He knew he had to change both public and congressional opinion, so in 1915 he organized a trip to the southern Sierra Nevada for several influential businessmen, magazine editors, and members of Congress. Mather's public relations succeeded, and in his second year on the job he was able to add several parks to the system, including Lassen Volcanic National Park in California, and Hawaii National Park. In 1917, Alaska's Mount McKinley was added to the system.

The stress of taking the reigns of the National Park Service (NPS) took its toll on Mather, and in 1917 he was forced to take 18 months off after suffering a nervous breakdown. In 1918 a rejuvenated Mather returned to work and promptly went about battling a measure aimed at allowing hydroelectric development in all of the national parks. The following year Mather added three more parks to the system—Zion (Utah), the Grand Canyon (Arizona), and Lafayette (now called Acadia, in Maine). In addition to his role in expanding the number of parks, Mather took the poorly organized and severely under funded NPS and turned it into the efficient organization that it is today. Mather continued working under three presidents and five secretaries of the Interior Department until a stroke forced him to retire in 1928.

John Muir (1838–1914)

Born in Dunbar, Scotland, John Muir immigrated to the United States with his family in 1849 at the age of 11. His family settled on a farm in Wisconsin where Muir's father put his children to work on the farm starting at 4 A.M., and working them up to 17 hours a day. In order to spend time by himself, Muir would wake up at 1 A.M. in order to work on inventions and read whatever books were available to him.

In 1860 Muir entered the University of Wisconsin, majoring in chemistry and geology, but left in 1863 without a degree. At the age of 29, he left on a 1,000-mile walking trip from Indiana to the Gulf of Mexico. From there he took a ship bound for South America and the Amazon, but on a stop in Cuba, Muir became extremely ill with malaria. Once he began to recover, he changed his plans and headed to California, arriving in 1868. He traveled through the Sierra Nevada and settled in the Yosemite Valley. He was the first person to explain the glacial origin of Yosemite Valley in an article published by the *New York Tribune* in 1871.

Muir was a pioneer in the area of wilderness preservation. He lobbied the federal government to create a national park system, taking his crusade all the way to Congress and the White House. In 1890 he persuaded Congress to pass the Yosemite National Park Bill, establishing Yosemite and Sequoia National Parks. In 1903 he spent four days camping out under the stars with Theodore Roosevelt, explaining his ideas on preservation to the president. In 1911 he submitted plans for a national park system to President Taft. This system was later established, after Muir's death, by Theodore Roosevelt in 1916.

One of Muir's greatest political battles came in the case of Yosemite's Hetch Hetchy Valley, which utilitarian conservationists, such as Director of the Forest Service Gifford Pinchot and Secretary of the Interior James Garfield, wanted to dam and flood as a water system for San Francisco. Muir and a handful of preservationists fought to block the dam to save the valley for posterity, but in the end the utilitarians won and the valley was eventually inundated. In 1882 Muir helped found the Sierra Club to bring western preservationists together into a united group of volunteers. Muir became the organization's first president; a post he held until his death in 1914.

Frederick Law Olmsted (1822–1903)

Born and raised in Hartford, Connecticut, Frederick Law Olmsted later moved to Staten Island, New York, where he took up the life of a farmer and operated a nursery and landscape business. In 1857, Olmsted won the job that would make him most famous: superintendent of New York City's future Central Park. Olmsted was responsible for designing and overseeing construction of the world's most ambitious city park to date, with over 800 acres of wooded hills, lakes, and meadows.

In 1863 Olmsted moved to California to manage the Mariposa mining estate. It was in the West that Olmsted first began to work for the preservation of America's wilderness, lobbying to place Yosemite Valley under state control as a wilderness park. Olmsted was appointed by the governor as the first commissioner of Yosemite State Park. While in this position he wrote a philosophical mandate, justifying the reasons for creating such parks as Yosemite. He wrote, "In permitting the sacrifice of anything that would be of the slightest value to future visitors ... we probably yield each case the interest of uncounted millions to the selfishness of a few individuals."[1] This statement was the first of its kind written in the United States. Olmsted wanted parks accessible for the enjoyment of all Americans, but wanted at the same time to keep roads and development to a minimum. Eventually Olmsted returned to New York and his landscape architecture career, where he worked on such projects as the National Zoo and the Capitol grounds in Washington, D.C.

Sigurd F. Olson (1899–1982)

Writer, philosopher, and preservationist are all words which describe Sigurd Olson. Olson was born in Chicago, but raised primarily in Sister Bay and Prentice, Wisconsin, on the shores of Lake Superior. He spent many days of his youth roaming through the neighboring forests. He studied agriculture at the University of Wisconsin, graduating in 1923, and received his master's degree in biology in 1931 from the University of Illinois.

During the 1930s, Sigurd Olson took a job teaching agriculture in a small mining town in northeastern Minnesota. Any time he could spare he spent on overnight canoe trips in the Quetico-Superior Wilderness Area and the Boundary Waters Canoe Area wilderness. When he decided that too much of the land was being opened to loggers, he began a lifelong fight to preserve it in its pristine state. He lectured and wrote articles, and was a consultant to the President's Quetico-Superior Committee (1947), a consultant to the Izaak Walton League of America (1947–1982), president of the National Parks Association (1953–1958), president of The Wilderness Society (1967–1971), and a member of the U.S. Department of the Interior's Advisory Committee on Conservation (1960–1966).

Sigurd Olson is perhaps best known for his writing, in which he inspired others to care for the natural world. He authored numerous books of wilderness ruminations, including *The Singing Wilderness* (1956), *Wilderness Days* (1972), *Reflections from the North Country* (1976), and *Time and Place* (1982). He wrote of the spiritual need inside mankind for the peace and solitude of nature; a craving dating to our primeval past. "The intangible values of wilderness are what really matter," he wrote, "the opportunity of knowing again what simplicity really means, the importance of the natural and the sense of oneness with the earth that inevitably comes with it."[2] In 1972 Northland College founded the Sigurd Olson Environmental Institute in his honor.

Gifford Pinchot (1865–1946)

During Theodore Roosevelt's conservation movement at the turn of the century, Gifford Pinchot was Roosevelt's right-hand man. As the head of the USFS during Roosevelt's administration, Pinchot was a driving force behind conservation reforms of the time.

Gifford Pinchot's childhood began far from the seclusion of the forests he would grow to love. Pinchot was born in New York City to a wealthy businessman, but his interest in nature and wilderness began at an early age. In 1885 Pinchot entered Yale University, and while he was most interested in forestry, not one college or university in the United States offered this field of study. In 1889, Pinchot traveled to Europe, where he met with several world-famous foresters, including the German Sir Dietrich

Brandis, who convinced Pinchot to attend the French Forest School in Nancy. Pinchot traveled extensively and studied at the school for one year.

Once back in the United States, Pinchot took a job supervising the 5,000-acre estate of George W. Vanderbilt, known as the Biltmore, in Asheville, North Carolina. Pinchot used this experience as a test to see how the forest could be managed to produce a constant yearly supply of wood products without irreparably damaging the forest itself. This exemplified Pinchot's philosophy of conservation, with utilitarian production managed by scientific monitoring. He ran into some opposition amongst other foresters who argued for preservation over conservation, but after leaving the Biltmore he went to work first for the Department of the Interior, and then the Department of Agriculture (USDA) where he headed the Division of Forestry.

Pinchot first met Theodore Roosevelt in 1899, and the two became close friends. When Roosevelt was elected president, he appointed Pinchot to head the newly created USFS. Together, Roosevelt, Pinchot, and Secretary of the Interior James R. Garfield ushered in a platform of wilderness conservation the likes of which the world had never seen. The U.S. federal government dramatically expanded the number of protected acres and began to manage grazing, logging, and other forest use.

John Wesley Powell (1834–1902)

Born in New York City but raised in Ohio, John Wesley Powell later became famous for his scientific exploration of the American West. Powell's father was a Methodist minister who encouraged his son to follow in his footsteps. Instead John became a teacher and studied at local colleges in Illinois. At the age of 27 he was elected the first secretary of the Illinois Natural History Society, but this position was interrupted by the outbreak of the Civil War. Powell joined the Illinois Volunteer Infantry and rose to the rank of major in the Union Army. At the battle of Shiloh, Powell was injured and lost his right arm, but still managed to return to active duty.

In 1865, with the war behind him, Powell became a natural history professor and returned to start a museum at the Illinois Natural History Society. It was at this point that he began what would become a life of travel and study in America's arid West. In 1867 Powell borrowed scientific instruments from the Smithsonian Institution and led an expedition to Colorado and the Rocky Mountains.

In 1871 Powell led his most daring expedition down the Colorado River through the entire length of the Grand Canyon (which Powell named). Faced with treacherous rapids and the frightening unknown, 3 of the 12 men left the group halfway through and hiked out of the canyon, only to be killed by American Indians once they reached the rim. The rest of the party continued on to become the first ever to successfully navigate that portion of the river.

Powell continued to study and write about this region of the United States for the rest of his life. He successfully lobbied Congress for money to conduct a survey of the region, and was responsible in a large part for the creation of the U.S. Geological Survey. In 1881 Powell became director of the Geological Survey, holding the post until 1894. In 1963 the construction of Glen Canyon dam created the nation's largest man-made reservoir, which was named Lake Powell in his honor.

William Kane Reilly (1940–)

William Reilly was born in Decatur, Illinois. He graduated from Yale University in 1962, and Harvard Law school in 1965, after which he went to work for a Chicago law firm. He went back to school to earn a master's degree in urban planning from Columbia University in 1971. Reilly is a devoted conservationist who worked on the President's Council on Environmental Quality from 1970–1972, the President's Conservation Foundation from 1973–1989, and the World Wildlife Federation from 1985–1989.

In 1989 Reilly was appointed to head the Environmental Protection Agency (EPA) by President George Bush. Reilly held the post for the duration of Bush's term, but it was an odd relationship. In public, President Bush stood up for his EPA chief, but behind closed doors they did not always agree. President Bush favored business in areas where Reilly favored the environment, and in many cases Reilly was forced to negotiate his agenda. At the 1992 Earth Summit in Brazil for example, news was leaked that Reilly was frustrated by a lack of cooperation from the president. In another case Reilly and Vice President Dan Quayle were at odds over wetlands protection, and the definition of wetland areas, but Reilly managed to work out a compromise.

Some environmentalists accused Reilly of giving in too easily to conservative interests, but most conservationists agreed that Reilly deserved praise for what he accomplished working with both environment and business advocates. He stopped a definition that would have slashed protected wetlands by 80 percent, halted plans for the Two Forks Dam on the Platte River in Colorado, and voted to strengthen the Clean Air Act to reduce acid rain. On the pro-business side, his wetlands compromise eased the existing definition of a wetland area, though not as much as Vice President Quayle wanted.

Theodore Roosevelt (1858–1919)

Theodore Roosevelt was one of the first U.S. presidents to bring conservation to the forefront of the national agenda. Roosevelt was born and raised in New York City. He graduated Phi Beta Kappa from Harvard University in 1880. In 1881 he was elected to the New York State Assembly. In 1895 Roosevelt became assistant Secretary of the Navy. During the Spanish–American War he became famous as the leader of the Rough Riders, a cavalry regiment. After the war he became governor of New York, and in 1900 was elected as vice president with President William McKinley. Six months later he became president when McKinley was assassinated.

As president, Roosevelt took a keen interest in preserving the environment. He traveled to California and met with John Muir, who was lobbying for a national park system. Roosevelt established the USFS in 1905. In 1908 he sponsored the first White House governors' conference, which helped publicize the conservation movement. In 1916 he initiated the NPS, and during his one and three-fourths terms in office, he added more than 125 million acres to the national forests.

Carl Schurz (1829–1906)

Carl Schurz probably has the most colorful background of any conservationist. Born in Germany, the liberal-minded Schurz became a student leader of an unsuccessful revolution in that country in 1848. His exploits included leading Prussian soldiers on chases through the streets and sewers of Bonn, and breaking one of his professors out of the Spandau prison. When the revolution failed, the 23-year-old Schurz and his wife moved to the United States, where he was hailed as a hero by the German-American community.

Once settled in his new land, Schurz became engulfed in the anti-slavery movement and began to campaign for presidential candidate Abraham Lincoln. Schurz was a powerful speaker who put his life into the campaign, and when Lincoln was elected, Schurz was appointed minister to Spain. He would not hold this position long, however, for when the Civil War broke out, he returned to the United States and became a general in the Union Army. He grew to be famous for parading up and down the front lines on his horse in the heat of battle, chomping on a cigar and yelling encouragement to his men.

After surviving the war, Schurz was named Secretary of the Interior by President Rutherford B. Hayes in 1877. Conservation was in its infancy at this time, but Schurz became the first Secretary of the Interior dedicated to preserving the nation's natural treasures, instead of simply divvying them up. This marked the beginning of a major shift in ideology, and for Schurz it was a difficult battle. He fought with Congress for laws to better protect the environment, and when he failed there, he did the best he could with existing laws. He sent agents out into the field to stop loggers from illegally cutting federal trees. Fines from these efforts resulted in a net gain for the program, but Congress called Schurz "un-American" and cut the budget for his agents. Though Schurz's gains for conservation were limited during his tenure, he is best remembered for being one of the first U.S. government officials to make conservation a priority issue, and his battles paved the way for future conservationists.

Stewart Udall (1920–)

Stewart Udall was born in Arizona, the son of Mormon missionaries. As a child he worked on his family's farm and did missionary work. During World War II he was a gunner on a B-24 bomber in Italy. He returned to the United States to attend the University of Arizona where he earned his law degree in 1948 and then went on to serve three terms as a United States congressman from Arizona's Second District. In 1961 Udall was appointed Secretary of the Interior by President John F. Kennedy. He held this position for eight years under both presidents Kennedy and Johnson.

As Secretary of the Interior, Stewart Udall made some of the greatest gains for conservation of anyone in that post. While in office, Udall helped add 59 areas to the National Park System, including six national seashores and four national parks, and created the first national system of wild and scenic rivers. He also oversaw the signing into law of the Wilderness Act in 1964.

In 1963, Udall worked 16-hour days to perform his duties and write the book *The Quiet Crisis,* a history of land reform to which President Kennedy contributed the introduction. Udall also wrote *1976: Agenda for Tomorrow* (1968), linking environmental and social changes.

James Gaius Watt (1938–   )

Former Secretary of the Interior James Watt was the most controversial figure to hold that post to date. Watt was born and raised in Lusk, Wyoming, and educated at the University of Wyoming, where he received a bachelor's degree in 1960 and a law degree in 1962. After law school he moved to Washington, D.C., where he held a number of government positions, from senatorial aide to Alan Simpson (R-WY), to lobbyist for the Chamber of Commerce, to director of the Bureau of Outdoor Recreation. In 1977 he became the chief legal advisor for the Mountain States Legal Foundation, a law firm which specializes in helping property owners challenge environmental laws.

In 1980 President Ronald Reagan first offered the Secretary of Interior position to Wyoming Senator Clifford Hansen. Hansen declined and Watt was then named to the post. The controversy surrounding Watt came from his attitude toward wilderness areas. As a born-again Christian, Watt believed strongly in the second coming of Christ, as he explained to a Congressional panel. For this reason he saw no point in protecting the wilderness for future generations, and worked to open protected areas for extraction of resources, attempting to offer oil, gas, and mineral leases in wilderness areas, national parklands, and offshore tracts, despite strong congressional opposition. During his first year as secretary he doubled the number of leases extended annually in unprotected areas.

Even President Reagan joked about Watt wanting to strip mine the presidential rose garden. Over one million Americans signed a petition demanding his removal from office, and this public outcry led to Watt's firing in 1983. He was succeeded by William P. Clark.

Howard Zahniser (1906–1964)

When Howard Zahniser was a young boy growing up in Pennsylvania's Allegheny River Valley, he had a difficult time seeing the beauty of the wilderness surrounding him. Zahniser had poor vision, and his family could not afford an optometrist, but when he finally did get a pair of glasses, in his teens, says his son, "he jumped fences and ran through fields, marveling at how much there was to see and how distinctly beautiful it was."[3] It was still several decades, however, before Zahniser dedicated himself to wilderness conservation.

Howard Zahniser went to Greenville College in Illinois, worked as a reporter at the Pittsburgh Press after graduating, and then worked as a high school English teacher from 1930 to 1945. He was 39 years old when he decided to quit teaching and accept a job as executive secretary of The Wilderness Society at half of his teaching salary. Howard Zahniser was not a highly charismatic leader, but he was an excellent writer, and dedicated to his cause. In a post-war America more concerned with economic growth than nature preservation, Zahniser made two monumental legislative gains. The first involved a battle over a proposed dam that would have flooded the 200,000-acre Dinosaur National Monument on the Colorado–Utah border. Zahniser, along with the Sierra Club's David Brower, rallied environmentalists and enlisted the help of senators Richard Neuberger and Hubert Humphrey to fight the proposal, and after a heated battle in both the House of Representatives and the Senate, the plan was eventually defeated in 1956.

Zahniser's greatest contribution to conservation, and the one for which he is most remembered, was his work on the Wilderness Act. Zahniser wrote the first version of the bill, which provided for a National Wilderness Preservation System, and had it introduced to Congress in 1956 by Senator Hubert Humphrey and Representative John P. Saylor. The bill was eventually rewritten 66 times over 8 years, but Zahniser stuck with the fight, attending every public hearing on the issue until he died of a heart attack in 1964, just weeks before the Wilderness Act passed into law.

Notes

1. Douglas H. Strong. *Dreamers and Defenders: American Conservationists* (Lincoln: University of Nebraska Press, 1988), p. 24.

2. Jim Dale Vickery. *Wilderness Visionaries* (Merrillville, Indiana: ICS Books, 1986), p. 194.

3. Peter Wild. *Pioneer Conservationists of Eastern America* (Missoula, MT: Mountain Press, 1986), p. 153.

4

Statistics, Quotations, and Legislation

THIS CHAPTER IS DIVIDED INTO three sections. Statistics provides data on the federal and state lands, location of wilderness, endangered species, and wetlands degradation. Quotations cites viewpoints of people on both sides of the preservation debate in an attempt to give a balanced look and the range of emotions behind some of the issues. Legislation summarizes and provides excerpts from some of the most influential laws affecting wilderness and the environment today.

Statistics

Most wilderness in the United States is located on federally owned lands. As mentioned in Chapter 1, these lands are divided among several different agencies, with most of the wilderness administered by the National Park Service (NPS), United States Fish and Wildlife Service (USFWS), Bureau of Land Management (BLM), and United States Forest Service (USFS). Table 4.1 and Figure 4.1 illustrate the division of these lands by acre.

These federal lands contain 484 national wildlife refuges, 589 wilderness areas, 156 national forests, and 368 units of the park

TABLE 4.1

Federal Land Totals	
Bureau of Land Management	265,449,727
Park Service	83,655,217
Fish and Wildlife Service	90,970,774
Forest Service	191,453,345
Bureau of Indian Affairs	2,700,000
Bureau of Reclamation	5,700,000
Defense Department	26,000,000
Miscellaneous	1,900,000
Total	667,829,063 acres

FIGURE 4.1
Federal Land Totals by Administering Agency

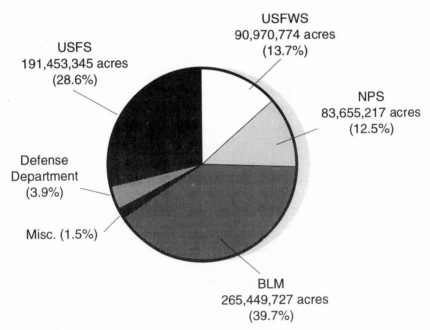

Total land: 668 million acres

system, including 53 national parks. Nearly 104 million acres of designated wilderness is located within these areas, as demonstrated in Figure 4.2.

Funds to operate the federal land agencies come from taxpayer dollars primarily. Congress appropriates the most money to

FIGURE 4.2
Designated Wilderness on the Federal Lands

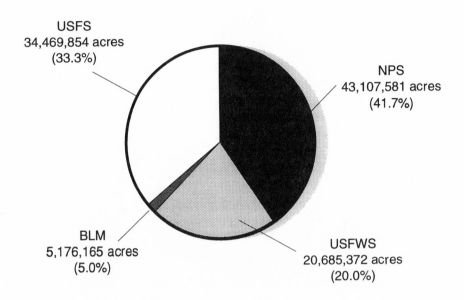

USFS
34,469,854 acres
(33.3%)

NPS
43,107,581 acres
(41.7%)

BLM
5,176,165 acres
(5.0%)

USFWS
20,685,372 acres
(20.0%)

Total Designated Wilderness: 103,438,972 acres

the USFS, followed by the NPS (Figure 4.3). In addition to these funds, the USFS generates approximately $1 billion per year through timber sales and other resources, for a total 1993 budget of $3.3 billion. The BLM also adds to their budget through sales of timber and land, grazing fees, and gas and oil leases. In 1992 these receipts totaled just under $257 million.

The national forests and national parks receive approximately the same number of recreation visitors each year, at more than 270 million each (Figure 4.4). The most visited unit in the National Park System is the Blue Ridge Highway, with 17,561,400 recreation visitors in 1992. The least visited unit is Aniakchak National Memorial in Alaska, with only 1,600 visitors in 1992. The most visited national park is Great Smoky Mountains National Park, with 8,931,700 recreational visitors in 1992. All in all, 17 national parks had more than 1 million recreational visitors each in that year (Table 4.2).

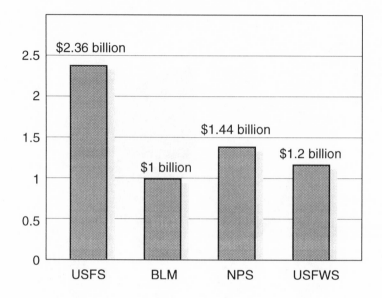

FIGURE 4.3
Federal Land Agency Appropriations, 1993

TABLE 4.2
Most Visited National Parks, 1992

National Park	Recreational Visitors
Great Smoky Mountains	8,931,700
Grand Canyon	4,203,500
Yosemite	3,819,500
Yellowstone	3,144,400
Olympic	3,030,200
Rocky Mountain	2,788,900
Mammoth Cave	2,392,900
Zion	2,390,600
Glacier	2,199,800
Shenandoah	1,822,200
Grand Teton	1,744,600
Mount Rainier	1,522,100
Hot Springs	1,504,000
Haleakala	1,183,100
Hawaii Volcanoes	1,151,700

FIGURE 4.4
Annual Recreation Visits to the Federal Lands, 1992

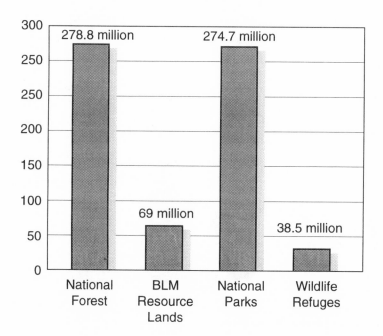

Alaska contains more federal land and wilderness than any state by far, with 240 million out of 668 million total acres of federal land, and 57.4 million out of 103.4 million acres of designated wilderness (Figures 4.5 and 4.7). It has eight national parks, two national monuments, ten national preserves, and one wild and scenic river .

Within the United States there are currently 278 native animals and 281 native plants on the endangered species list. An additional 99 animals and 71 plants are listed as threatened, for a total of 729 endangered and threatened species. Outside the United States, 490 species are listed as endangered and 39 as threatened, for a grand total of 1,258 species listed worldwide. Approximately 3,600 species have been nominated to be listed worldwide, but are awaiting scientific review. Listed species within the United States include the grizzly bear, bald eagle, peregrine falcon, gray wolf, California condor, Florida panther, northern

FIGURE 4.5
Federal Land: Continental United States vs. Alaska

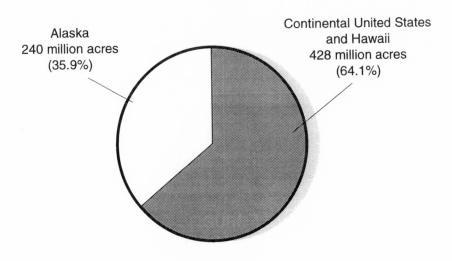

Alaska
240 million acres
(35.9%)

Continental United States
and Hawaii
428 million acres
(64.1%)

Total federal land: 668 million acres

spotted owl, black-footed ferret, and several species of sea turtles and salmon.

The greatest cause of species loss in the United States and around the world is the loss of specific habitats. Most old-growth forest in the United States has been already cut, including over 90 percent of the Pacific Northwest's original 25 million acres. The prairies have been converted to farmland, with only one one hundredth of one percent of the original tallgrass prairie remaining today. One habitat that still remains in many regions is the wetland. Wetlands support a large variety of life, including aquatic plants and animals, and many bird species. Migrating birds depend on wetlands as they stop to rest or stay for extended periods during their long journeys. Today, wetlands are among the most threatened of natural habitats, as they have been reduced in the continental United States by 53 percent (Table 4.4). Six states have had losses approaching 90 percent or more.

FIGURE 4.6
Park Service Lands: Continental United States vs. Alaska

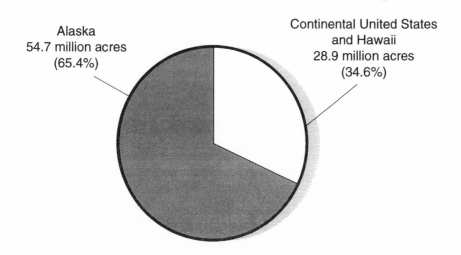

Alaska
54.7 million acres
(65.4%)

Continental United States
and Hawaii
28.9 million acres
(34.6%)

Park Service total: 83,655,217 acres

FIGURE 4.7
Designated Wilderness: Continental United States vs. Alaska

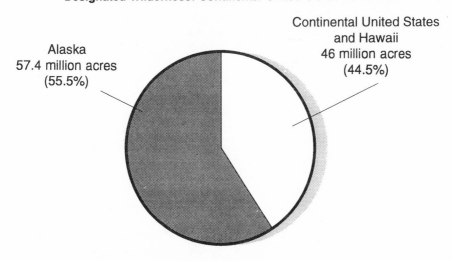

Alaska
57.4 million acres
(55.5%)

Continental United States
and Hawaii
46 million acres
(44.5%)

Total wilderness: 103.4 million acres

TABLE 4.3
U.S. Endangered and Threatened Species

Group	Endangered	Threatened	Total
Mammals	56	9	65
Birds	73	12	85
Reptiles	16	18	34
Amphibians	6	5	11
Fishes	55	36	91
Snails	7	6	13
Clams	40	2	42
Crustaceans	8	2	10
Insects	14	9	23
Arachnids	3	0	3
Subtotal	278	99	377
Plants	281	71	352
Total	559	170	729

Source: U.S. Fish and Wildlife Service, as of 30 September 1992.

TABLE 4.4
Wetland Losses in the United States 1780s to 1980s

State	Total Acres	Estimated Wetlands 1780s	Estimated Wetlands 1980s	Percent Lost
AL	33,029,760	7,567,600	3,783,800	50%
AZ	72,901,760	931,000	600,000	36%
AR	33,986,560	9,848,600	2,763,600	72%
CA	101,563,520	5,000,000	454,000	91%
CO	66,718,720	2,000,000	1,000,000	50%
CT	3,205,760	670,000	172,500	74%
DE	1,316,480	479,785	223,000	54%
FL	37,478,400	20,325,013	11,038,300	46%
GA	37,680,640	6,843,200	5,298,200	23%
ID	53,470,080	877,000	385,700	56%
IL	36,096,000	8,212,000	1,254,500	85%
IN	23,226,240	5,600,000	750,633	87%
IA	36,025,600	4,000,000	421,900	89%
KS	52,648,960	841,000	435,400	48%
KY	25,852,800	1,566,000	300,000	81%
LA	31,054,720	16,194,500	8,784,200	46%
ME	21,257,600	6,460,000	5,199,200	20%
MD	6,769,280	1,650,000	440,000	73%
MA	5,284,480	818,000	588,486	28%
MI	37,258,240	11,200,000	5,583,400	50%
MN	53,803,520	15,070,000	8,700,000	42%

State	Total Acres	Estimated Wetlands 1780s	Estimated Wetlands 1980s	Percent Lost
MS	30,538,240	9,872,000	4,067,000	59%
MO	44,844,000	4,844,000	643,000	87%
MT	94,168,320	1,147,000	840,300	27%
NE	49,425,280	2,910,500	1,905,500	35%
NV	70,745,600	487,350	236,350	52%
NH	5,954,560	220,000	200,000	9%
NJ	5,015,040	1,500,000	915,960	39%
NM	77,866,240	720,000	481,900	33%
NY	31,728,640	2,562,000	1,025,000	60%
NC	33,655,040	11,089,500	5,689,500	49%
ND	45,225,600	4,927,500	2,490,000	49%
OH	26,382,080	5,000,000	482,800	90%
OK	44,748,160	2,842,600	949,700	67%
OR	62,067,840	2,262,000	1,393,900	38%
PA	29,013,120	1,127,000	499,014	56%
RI	776,960	102,690	65,154	37%
SC	19,875,200	6,414,000	4,659,000	27%
SD	49,310,080	2,735,100	1,780,000	35%
TN	27,036,160	1,937,000	787,000	59%
TX	171,096,960	15,999,700	7,612,412	52%
UT	54,346,240	802,000	558,000	30%
VT	6,149,760	341,000	220,000	35%
VA	26,122,880	1,849,000	1,074,613	42%
WA	43,642,880	1,350,000	938,000	31%
WV	15,475,840	134,000	102,000	24%
WI	35,938,560	9,800,000	5,331,392	46%
WY	62,644,960	2,000,000	1,250,000	38%
SUBTOTAL	1,934,198,400	221,129,638	104,374,314	53%
ALASKA	375,303,680	170,200,000	170,000,000	0.1%
HAWAII	4,115,200	58,800	51,800	12%
TOTAL	2,313,617,280	391,388,438	274,426,114	30%

Source: Dahl, T. E. *Wetlands Losses in the United States 1780's to 1980's.* Washington, DC: U.S. Department of the Interior, Fish and Wildlife Service, 1990, p. 6.

Quotations

The following series of quotations and passages concerning wilderness is intended to provide a feel for the opinion and emotion on both sides of the issues.

We need to find a thoughtful, humane balance in the American West which says we can live together with the natural environment. We don't have to have these endless confrontational wars which are all premised by "it's us versus them." If its jobs versus the environment, one side gains only as the other loses.

Bruce Babbitt, 1993

If we are going to make the Endangered Species Act work, we've got to look at the whole ecosystem. We have to use strong, comprehensive science to survey the whole place in advance of the crisis, get the information, see it whole, see not one, but many species, and then—while you still have the flexibility—work something out.

Bruce Babbitt, 1993

If we are to succeed as inhabitants of a world increasingly transformed by technology, we need to reassess our attitudes toward the natural world on which our technology intrudes. For in the eager search for the benefits of modern science and technology, we have become enticed into a nearly fatal illusion: that we have at last escaped from the dependence of man on the rest of nature.

The truth is tragically different. We have become, not less dependent on the balance of nature, but more dependent on it. Modern technology has so stressed the web of processes in the living environment at its most vulnerable points that there is little leeway left in the system. And time is short; we must begin, now, to learn how to make our technological power conform to the more powerful constraints of the living environment.

Barry Commoner, 1969

I am awfully glad my friends from California and elsewhere are getting tired of this conservation hobby, because, Mr. Chairman, I think it is one of the delusions of the age in which we live.

I sympathize with my friends in California who want to take a part of the public domain, now, notwithstanding all their declamations for conservation of resources. I am willing to let them have it. I am willing to let them have it when they take it in California and San Francisco for the public good.

That is what the great resources of this country are for. They are for the American people. I want them to open the coal mines in Alaska. I want them to open the reservations in this country. I am not for reservations and parks. I would have the great timber and mineral and coal resources of this country opened to the people, and I only want to say, Mr. Chairman, that your Pinchots and your conservationists generally are theorists who are not, in my humble judgment, making propaganda in the interest of the American people.

Let California have it, and let Alaska open her coal mines. God almighty has located the resources of this country in such a form as that His children will not use them in disproportion, and your Pinchots will not be able to controvert and circumvent the laws of God Almighty.

Representative Martin Dies, of Texas, during a Congressional debate on the proposed Hetch-Hetchy reservoir, 1913

Conservation is not a living ethic in many nations. As a way of life it probably has more articulate advocates in the United States than anywhere else, yet at home more ground is lost annually than gained. Pressures of commercial interests, of motorized recreationists, of our mounting population, threaten to over-run the meager wilderness areas left, fill them with the debris of civilization, and leave only alpine areas in primitive condition.

William O. Douglas, 1965, from *A Wilderness Bill of Rights*

I propose to speak for those exiles in sin who hold that a large part of the present "conservation" movement is unadulterated humbug. That the modern Jeremiahs are as sincere as was the older one, I do not question. But I count their prophecies to be baseless vaporings, and their vaunted remedy worse than the fancied disease. I am one who can see no warrant of law, of justice, nor of necessity for that wholesale reversal of our traditional policy which the advocates of "conservation" demand. I am one who does not shiver for the future at the sight of a load of coal, nor view a steel-mill as the arch-robber of posterity. I am one who does not believe in a power trust, past, present or to come; and who, if he were a capitalist seeking to form such a trust, would ask nothing better than just the present conservation scheme to help him. I believe that a government bureau is the worst imaginable landlord; and that its essential nature is not changed by giving it a high-sounding name, and decking it with home-made haloes. I hold that the present forest policy ceases to be a nuisance only when it becomes a curse.

George L. Knapp, 1910

How many of those whole-hearted conservationists who berate the past generation for its shortsightedness in the use of natural resources have stopped to ask themselves for what new evils the next generation will berate us?

Has it ever occurred to us that we may unknowingly be just as short-sighted as our forefathers in assuming certain things to be inexhaustible, and becoming conscious of our error only after they have practically disappeared?

Today it is hard for us to understand why our prodigious waste of standing timber was allowed to go on—why the exhaustion of the supply was not earlier foreseen. Some even impute to the waters a certain moral turpitude. We forget that for many generations the standing timber of America was in fact an encumbrance or even an enemy, and that the nation was simply unconscious of the possibility of its becoming exhausted. In fact, our tendency is not to call things resources until the supply runs short. When the end of the supply is in sight we "discover" that the thing is valuable.

Aldo Leopold, 1925

It is inconceivable to me that an ethical relation to land can exist without love, respect, and admiration for land, and a high regard for its value. By value, I of course mean something far broader than mere economic value; I mean value in the philosophical sense.

Aldo Leopold, 1949

Wilderness skeptics in almost all arguments raise the question: "Why should we set aside a vast area for the enjoyment of a few hundred people when roads would make that area available for half a million? Aren't we obligated to consider what will bring the greatest good to the greatest number?"

The doctrine of the greatest good to the greatest number does not mean that this laudable relationship has to take place on every acre. If it did, we would be forced to change our metropolitan art galleries into metropolitan bowling alleys. Our state universities, which are used by a minor fraction of the population, would be converted into state circuses where hundreds could be exhilarated for every one person who may be either exhilarated or depressed now. The Library of Congress would become a national hot dog stand, and the Supreme Court building would be converted into a gigantic garage where it could house a thousand people's autos instead of Nine Gentlemen of the Law.

Ridiculous as all of this sounds, it is no more ridiculous than the notion that every acre of land devoted to outdoor recreation should be administered in a way that will give the maximum volume of use. Quality as well as quantity must enter into any evaluation of competing types of recreation, because one really deep experience may be worth an infinite number of ordinary experiences. Therefore, it is preposterous to hold that the objective of outdoor recreational planning should be to enable the maximum number of people to enjoy every beautiful bit of the outdoors.

Robert Marshall, 1937

The Universe of the Wilderness is disappearing like a snowbank on a south-facing slope on a warm June day.

Robert Marshall

There is just one hope of repulsing the tyrannical ambition of civilization to conquer every niche on the whole earth. That hope is the organization of spirited people who will fight for the freedom of the wilderness.

Robert Marshall

The tendency nowadays to wander in the wilderness is delightful to see. Thousands of tired nerve-shaken, over-civilized people are beginning to find out that going to the mountains is going home; that wildness is a necessity; and that mountain parks and reservations are useful not only as fountains of timber and irrigating rivers, but as fountains of life. Awakening from the stupefying effects of the vice of over-industry and the deadly apathy of luxury, they are trying as best they can to mix and enrich their own little ongoings with those of Nature, and to get rid of rust and disease. Briskly venturing and roaming, some are washing off sins and cobweb cares of the devil's spinning in all-day storms on mountains; sauntering in rosiny pinewoods or in gentian meadows, brushing through chaparral, bending down and parting sweet, flowery sprays; tracing rivers to their sources, getting in touch with the nerves of Mother Earth; jumping from rock to rock, feeling the life of them, learning the songs of them, panting in whole-souled exercise and rejoicing in deep, long-drawn breaths of pure wildness. This is fine and natural and full of promise. And so also is the growing interest in the care and preservation of forests and wild places in general, and in the half-wild parks and gardens or towns. Even the scenery habit in its most artificial forms, mixed with spectacles, silliness, and kodaks; its devotees arrayed more gorgeously than scarlet tanagers, frightening wild game with red umbrellas—even this is encouraging, and may well be regarded as a hopeful sign of the times.

All the Western mountains are still rich in wildness, and by means of good roads are being brought nearer civilization every year. To the sane and free it will hardly seem necessary to cross the continent in search of wild beauty, however easy the way, for they find it in abundance wherever they chance to be. Like Thoreau they see forests in orchards and patches of huckleberry brush, and oceans in ponds and drops of dew. Few in these hot, dim, frictiony times are quite sane or free; choked with care like clocks full of dust, laboriously doing so much good and making so much money,—or so little,—they are no longer good themselves.

Man, too, is making many far-reaching changes. This most influential half animal, half angel is rapidly multiplying and spreading, covering the seas and lakes with ships, the land with huts, hotels, cathedrals, and clustered city shops and homes, so that soon, it would seem, we may have to go farther than Nansen to find a good sound solitude. None of Natures's landscapes are ugly so long as they are wild; and much, we can say comfortingly, must always be in great part wild, particularly the sea and the sky, the floods of light from the stars, and the warm, unspoilable heart of the earth, infinitely beautiful, though only dimly visible to the eye of imagination. The geysers, too, spouting from the hot underworld; the steady, long-lasting glaciers on the mountains, obedient only to the sun; Yosemite domes and the tremendous grandeur of rock canyons and mountains in general,—these must always be wild, for man can change them and mar them hardly more than can the butterflies that hover above them.

John Muir, 1898

The gentleman who introduced me did me the honor of mentioning the attention I devoted to this subject years ago as Secretary of the

Interior. When I entered upon that important office, having the public lands in charge, I considered it my first duty to look around me and to study the problems I had to deal with. Doing so I observed all the wanton waste and devastation I have described. I observed the notion that the public forests were everybody's property, to be taken and used or wasted as anybody pleased, everywhere in full operation. I observed enterprising timber thieves not merely stealing trees, but stealing whole forests. I observed hundreds of sawmills in full blast, devoted exclusively to the sawing up of timber stolen from the public lands.

I observed a most lively export trade going on from Gulf ports as well as Pacific ports, with fleets of vessels employed in carrying timber stolen from the public lands to be sold in foreign countries, immense tracts being devastated that some robbers might fill their pockets.

I thought this sort of stealing was wrong, in this country no less than elsewhere. Moreover, it was against the spirit and letter of the law. I, therefore, deemed it my duty to arrest the audacious and destructive robbery. Not that I had intended to prevent the settler and the miner from taking from the public lands what they needed for their cabins, their fields, or their mining shafts; but I deemed it my duty to stop at least the commercial depredations upon the property of the people. And to that end I used my best endeavors and means at my disposal, scanty as they were.

What was the result? No sooner did my attempts in that direction become known, than I was pelted with telegraphic dispatches from the regions most concerned, indignantly inquiring what it meant that an officer of the Government dared to interfere with the legitimate business of the country! Members of Congress came down upon me, some with wrath in their eyes, others pleading in a milder way, but all solemnly protesting against my disturbing their constituents in this peculiar pursuit of happiness. I persevered in the performance of my plain duty. But when I set forth my doing in my annual report and asked Congress for a rational forestry legislation, you should have witnessed the sneers at the outlandish notions of this "foreigner" in the Interior Department; notions that, as was said, might do for a picayunish German principality, but were altogether contemptible when applied to this great and free country of ours.

Carl Schurz, 1889, from a speech titled "The Need of a Rational Forest Policy," delivered to the American Forestry Association and the Pennsylvania Forestry Association

In Europe people talk a great deal of the wilds of America, but the Americans themselves never think about them; they are insensible to

the wonders of inanimate nature and they may be said not to per-
ceive the mighty forests that surround them till they fall beneath the
hatchet.

Alexis de Tocqueville, 1832

If we do not allow the private marketplace to go in and develop
energy sources in a systematic, methodical, and environmentally
sensitive way, we will create such a political and economic crisis that
Washington will nationalize the industries and attack our energy-rich
West in such a manner as to destroy the ecology, primarily because it
must get to that energy to heat the homes of the Northeast and keep
the wheels of industry going in the Midwest.

James Watt, 1981

The trumpeting voice of the wilderness lover is heard at great
distances these days. He is apt to be a perfectly decent person, if
hysterical. And the causes which excite him so are generally worthy.
Who can really find a harsh word for him as he strives to save Lake
Erie from the sewers of Cleveland, save the redwoods from the
California highway engineers, save the giant rhinoceros from the
Somali tribesmen who kill those noble beasts to powder their horns
into what they fondly imagine is a wonder-working aphrodisiac?
     Worthy causes, indeed, but why do those who espouse them have
to be so shrill and intolerant and sanctimonious? What right do they
have to insinuate that anyone who does not share their passion for
the whooping crane is a Philistine and a slob? From the gibberish
they talk, you would think the only way to save the bald eagle is to
dethrone human reason.
     I would like to ask what seems to me an eminently reasonable
question: *Why shouldn't we spoil the wilderness?*
     Have these people ever stopped to think what the wilderness
is? It is precisely what man has been fighting against since he began
his painful, awkward climb to civilization. It is the dark, the formless,
the terrible, the old chaos which our fathers pushed back, which sur-
rounds us yet, which will engulf us all in the end. It is held at bay by
constant vigilance, and when the vigilance slackens it swoops down
for a melodramatic revenge, as when the jungle took over Chichen
Itza in Yucatan, or lizards took over Jamshid's courtyard in Persia. It
lurks in our own hearts, where it breeds wars and oppression and
crimes. Spoil it! Don't you wish we could?
     Of course, when the propagandists talk about unspoiled
wilderness, they don't mean anything of that sort. What they mean by

wilderness is a kind of grandiose picnic ground, in the Temperate Zone, where the going is rough enough to be challenging but not literally murderous, where hearty folk like Supreme Court Justice Douglas and Interior Secretary Udall can hike and hobble through spectacular scenery, with a helicopter hovering in the dirty old civilized background in case a real emergency comes up.

Well, the judge and the Secretary and their compeers are all estimable people, and there is no reason why they should not be able to satisfy their urge for primitive living. We ought to recognize, however, that other people have equally strong and often equally legitimate urges to build roads, dig mines, plow up virgin land, erect cities. Such people used to be called pioneers; now they are apt to be called louts. At all events, we are faced with sets of conflicting drives, and it is up to us to make a rational choice among them.

The trouble is, it is difficult to make a rational choice when one of the parties insists on wrapping all its discourse in a vile metaphysical fog.

Robert Wernick, 1965

Legislation

Congress has passed laws that have affected the American wilderness ever since the first public lands were set aside in 1780. These laws have evolved over the years, reflecting the changing attitudes Americans have toward their open spaces (see Chapter 2, page 37). Some of the most influential laws regarding wilderness in the United States today include the Alaska National Interest Lands Conservation Act (ANILCA), Clean Air Act, Endangered Species Act, Federal Water Pollution Control Act (Clean Water Act), National Environmental Policy Act (NEPA), and the Wilderness Act. Following are discussions of and excerpts from these acts.

Alaska National Interest Lands Conservation Act (ANILCA)

Out of a total of 668 million acres of federal land, 240 million are located in Alaska. The Alaska National Interest Lands Conservation Act (ANILCA) designated 56.5 million of those acres as wilderness areas (nearly two-thirds of the United States total) and set sweeping guidelines for managing the total federal acreage.

The act divides much of Alaska's federal lands into national forests, wildlife refuges, national parks, reserves, and wild and scenic rivers. Thirteen national park units totaling 42.8 million acres were created by the act, including seven national parks: Gates of the Arctic, Glacier Bay, Katmai, Kenai Fjords, Kobuk Valley, Lake Clark, and Wrangell–St. Elias. A 1981 House of Representatives report on the act describes it as follows:

> The principal purpose of H.R. 39, as reported by the Committee, is to designate approximately 105-million acres of Federal land in Alaska for the protection of their resource values under permanent Federal ownership and management. The bill more than doubles the size of the National Park System and the National Wildlife Refuge System. It triples the size of the National Wilderness Preservation System. It virtually completes the public land allocation process in Alaska which began with the Statehood Act of 1958 . . .

The Statement of Purpose of the act begins:

(a) Establishment of units
    In order to preserve for the benefit, use, education, and inspiration of present and future generations certain lands and waters in the State of Alaska that contain nationally significant natural, scenic, historic, archeological, geological, scientific, wilderness, cultural, recreational, and wildlife values, the units described in the following titles are hereby established.

(b) Preservation and protection of scenic, geological, etc., values
    It is the intent of Congress in this Act to preserve unrivaled scenic and geological values associated with natural landscapes; to provide for the maintenance of sound populations of, and habitat for, wildlife species of inestimable value to the citizens of Alaska and the Nation, including those species dependent on vast relatively undeveloped areas; to preserve in their natural state extensive unaltered arctic tundra, boreal forest, and coastal rainforest ecosystems; to protect the resources related to subsistence needs; to protect and preserve historic and archeological sites, rivers, and lands, and to preserve wilderness resource values and related recreational opportunities including but not limited to hiking, canoeing, fishing, and sport hunting, within large arctic and subarctic wildlands and on freeflowing rivers; and to maintain opportunities for scientific research and undisturbed ecosystems.

Clean Air Act

The first legislation relating to air pollution was passed in 1955, but this only provided for federal research. The Clean Air Act of 1963 sent funds to individual states for use in cutting air pollution. Over the next 30 years, this act was amended four times, most recently in 1990. Each amendment requires stricter controls on emissions, which are monitored by the EPA.

The 1990 law regulates 189 hazardous metal and organic substances, such as lead, ozone, and carbon monoxide, that are emitted from oil refineries, chemical plants, and other industrial sources. It also regulates ozone-depleting chemicals which include chloroflourocarbons, halons, methyl chloroform, carbon tetrachloride, and hydrochloroflourocarbons.

Automobile manufacturers are required to reduce tailpipe emissions of hydrocarbons by 35 percent and nitrogen oxides by 60 percent on all vehicles sold by 1996, and another 50 percent reduction of both by 2003, unless the EPA finds the standards not necessary, technologically feasible, or cost effective. In addition, 300,000 clean fuel vehicles that run on methanol, ethanol, natural gas, or reformulated gasoline, must be sold annually in California by 1999. The law requires that the state of California create a program to meet these requirements. In lieu of a program requiring clean fuel vehicles, the California Air Reserves Board has mandated that 2 percent (or 40,000) of all cars sold in the state by 1998 must be electric, with increases to 5 percent by 2001, and 10 percent by 2003. General Motors, Ford, Chrysler, Honda, Nissan, Mazda and Toyota are required to comply by 1998. Smaller manufacturers must begin to comply by 2003. Any company that fails to meet the 2 percent quota in sales will lose their certification to sell cars in California.

Following is an excerpt from the Clean Air Act of 1967, explaining the reasons for the passage of the act.

FINDINGS AND PURPOSES

Sec. 101. (a) The Congress finds—

(1) that the predominant part of the Nation's population is located in its rapidly expanding metropolitan and other urban areas, which generally cross the boundary lines of local jurisdictions and often extend into two or more States;

(2) that the growth in the amount and complexity of air pollution brought about by urbanization, industrial development, and the increasing use of motor vehicles, has resulted in injury to agricultural crops and livestock, damage to and the deterioration of property, and hazards to air and ground transportation;

(3) that the prevention and control of air pollution at its source is the primary responsibility of States and local governments; and

(4) that Federal financial assistance and leadership is essential for the development of cooperative Federal, State, regional, and local programs to prevent and control air pollution.

(b) The purposes of this title are—

(1) to protect and enhance the quality of the Nation's air resources so as to promote the public health and welfare and the productive capacity of its population;

(2) to initiate and accelerate a national research and development program to achieve the prevention and control of air pollution;

(3) to provide technical and financial assistance to State and local governments in connection with the development and execution of their air pollution prevention and control programs; and

(4) to encourage and assist the development and operation of regional air pollution control programs.

Endangered Species Act

The USFWS and the National Marine Fisheries Service began keeping the list of endangered species after the first law to protect such species was passed in 1967. This act was amended in 1969, but still provided little in the way of actual protection. The law now known as the Endangered Species Act was passed in 1973 in an effort to protect these plants and animals. To achieve this end, the act requires all federal agencies to weigh their policies in regards to those species on the list. Development projects, logging, farming, or even military maneuvers can be halted if they are destroying endangered species or their habitat. Federal agencies can set aside or acquire land "by purchase, donation, or otherwise," or develop captive breeding programs. The act also makes it illegal to hunt, sell, purchase, transport, or possess any endangered species. The title and findings, purposes and policy section in this act read:

An Act

To provide for the conservation of endangered and threatened species of fish, wildlife, and plants, and for other purposes.

Be it enacted by the Senate and House of Representatives of the United States of America in Congress assembled. That this Act may be cited as the "Endangered Species Act of 1973".

FINDINGS, PURPOSES, AND POLICY

SEC. 2. (a) FINDINGS—The Congress finds and declares that—

(1) various species of fish, wildlife, and plants in the United States have been rendered extinct as a consequence of economic growth and development untempered by adequate concern and conservation;

(2) other species of fish, wildlife, and plants have been so depleted in numbers that they are in danger of or threatened with extinction;

(3) these species of fish, wildlife, and plants are of aesthetic, ecological, educational, historical, recreational, and scientific value to the Nation and its people;

(4) the United States has pledged itself as a sovereign state in the international community to conserve to the extent practicable the various species of fish or wildlife and plants facing extinction, pursuant to—

(A) migratory bird treaties with Canada and Mexico;

(B) the Migratory and Endangered Bird Treaty with Japan;

(C) the Convention on nature Protection and Wildlife Preservation in the Western Hemisphere;

(D) the International Convention for the Northwest Atlantic Fisheries;

(E) the International Convention for the High Seas Fisheries of the North Pacific Ocean;

(F) the Convention on International Trade in Endangered Species of Wild Fauna and Flora; and

(G) other international agreements; and

(5) encouraging the States and other interested parties, through Federal financial assistance and a system of incentives, to develop and maintain conservation programs which meet national and international standards is a key to meeting the Nation's international commitments and to better safeguarding, for the benefit of all citizens, the Nation's heritage in fish, wildlife, and plants.

(b) PURPOSES—The purposes of this chapter are to provide a means whereby the ecosystems upon which endangered species and threatened species depend may be conserved, to provide a program for the conservation of such endangered species and threatened species, and to take such steps as may be appropriate to achieve the purposes of the treaties and conventions set forth in subsection (a) of this section.

(c) POLICY—

(1) It is further declared to be the policy of Congress that all Federal departments and agencies shall seek to conserve endangered species and threatened species and shall utilize their authorities in furtherance of the purposes of this chapter.

(2) It is further declared to be the policy of Congress that Federal agencies shall cooperate with State and local agencies to resolve water resource issues in concert with conservation of endangered species.

Federal Water Pollution Control Act

This law is better known as the Clean Water Act. As section 101 reads, "The objective of this Act is to restore and maintain the chemical, physical, and biological integrity of the Nation's waters." To achieve this goal, the act tightly regulates waste treatement management to halt the discharge of pollutants into "the navigable waters, waters of the contiguous zone, and the oceans."

Environmentalists have used this act as a tool to protect wetlands. Most wetland degradation comes when farmers or developers dump dirt or other material into a wetland in order to fill it in. Section 404 of the Clean Water Act requires a permit from the Army Corps of Engineers to dump fill material, and that permit can be denied if the discharge will adversely effect shellfish beds, fishery areas, or wildlife. The law was not written with a specific goal of protecting wetlands, and in fact, the term wetland is not even mentioned.

Since it is only illegal to fill in wetlands, some developers simply began to dig channels to drain them instead. Also, no authoritative definition of wetlands exists (see Chapter 1, p. 29–31). Overall, the legal protection for wetlands is slim, but President Clinton called for a new Clean Water Act in 1994, and proponents are expecting written protection aimed specifically at the nations remaining wetlands. Here is Section 404 of the 1972 Clean Water Act:

PERMITS FOR DREDGED OR FILL MATERIAL

SEC. 404. (a) The Secretary of the Army, acting through the Chief of Engineers, may issue permits, after notice and opportunity for public hearings for the discharge of dredged or fill material into navigable waters at specified disposal sites.

(b) Subject to subsection (c) of this section, each such disposal site shall be specified for each such permit by the Secretary of the Army (1) through the application of guidelines developed by the Administrator, in conjunction with the Secretary of the Army, which guidelines shall be based upon criteria comparable to the criteria applicable to the territorial seas, the contiguous zone, and the ocean under section 403(c), and (2) in any case where such guidelines under clause (1) alone would prohibit the specification of a site, through the application additionally of the economic impact of the site on navigation and anchorage.

(c) The Administrator is authorized to prohibit the specification (including the withdrawal of specification) of any defined area as a disposal site, and he is authorized to deny or restrict the use of any defined area for specification (including the withdrawal of specification) as a disposal site, whenever he determines, after notice and opportunity for public hearings, that the discharge of such materials into such area will have unacceptable adverse effect on municipal water supplies, shellfish beds and fishery areas (including spawning and breeding areas), wildlife, or recreational areas. Before making such determination, the Administrator shall consult with the Secretary of the Army. The Administrator shall set forth in writing and make public his findings and his reasons for making any determination under this subsection.

National Environmental Policy Act (NEPA)

NEPA requires all federal agencies to consider the environment in all of their activities. If an activity may have an impact on the environment, the agency is required to prepare and submit an environmental impact statement before any other action is taken. Following is a portion of this act:

A. The National Environmental Policy Act of 1969, As Amended

An Act to establish a national policy for the environment, to provide for the establishment of a Council on Environmental Quality, and for other purposes.

Be it enacted by the Senate and House of Representatives of the United States of America in Congress assembled, That this Act may be cited as the "National Environmental Policy Act of 1969."

PURPOSE

SEC. 2. The purposes of this Act are: To declare a national policy which will encourage productive and enjoyable harmony between man and his environment and biosphere and stimulate the health and welfare of man; to enrich the understanding of the ecological systems and natural resources important to the Nation; and to establish a Council on Environmental Quality.

TITLE 1
DECLARATION OF NATIONAL ENVIRONMENTAL POLICY

SEC. 101. (a) The Congress, recognizing the profound impact of man's activity on the interrelations of all components of the natural environment, particularly the profound influences of population growth, high density urbanization, industrial expansion, resource exploitation, and new and expanding technological advances and recognizing further the critical importance of restoring and maintaining environmental quality to the overall welfare and development of man, declares that it is the continuing policy of the Federal Government, in cooperation with State and local governments, and other concerned public and private organizations, to use all practicable means and measures, including financial and technical assistance, in a manner calculated to foster and promote the general welfare, to create and maintain conditions under which man and nature can exist in productive harmony, and fulfill the social, economic, and other requirements of present and future generations of Americans.

(b) In order to carry out the policy set forth in this Act, it is the continuing responsibility of the Federal Government to use all practicable means, consistent with other essential considerations of national policy, to improve and coordinate Federal plans, functions, programs, and resources to the end that the Nation may—

(1) fulfill the responsibilities of each generation as trustee of the environment for succeeding generations;

(2) assure for all Americans safe, healthful, productive, and aesthetically and culturally pleasing surroundings;

(3) attain the widest range of beneficial uses of the environment without degradation, risk to health or safety, or other undesirable and unintended consequences;

(4) preserve important historic, cultural, and natural aspects of our national heritage, and maintain, wherever possible, an environment which supports diversity, and variety of individual choice;

(5) achieve a balance between population and resource use which will permit high standards of living and a wide sharing of life's amenities; and

(6) enhance the quality of renewable resources and approach the maximum attainable recycling of depletable resources.

(c) The Congress recognizes that each person should enjoy a healthful environment and that each person has a responsibility to contribute to the preservation and enhancement of the environment.
SEC. 102 The Congress authorizes and directs that, to the fullest extent possible: (1) the policies, regulations, and public laws of the United States shall be interpreted and administered in accordance with the policies set forth in this Act, and (2) all agencies of the Federal Government shall—

(A) Utilize a systematic, interdisciplinary approach which will insure the integrated use of the natural and social sciences and the environmental design arts in planning and in decisionmaking which may have an impact on man's environment;

(B) Identify and develop methods and procedures, in consultation with the Council on Environmental Quality established by title II of this Act, which will insure that presently unquantified environmental amenities and values may be given appropriate consideration in decision making along with economic and technical considerations;

(C) Include in every recommendation or report on proposals for legislation and other major Federal actions significantly affecting the quality of the human environment, a detailed statement by the responsible official on—

(i) The environmental impact of the proposed action,

(ii) Any adverse environmental effects which cannot be avoided should the proposal be implemented,

(iii) Alternatives to the proposed action,

(iv) The relationship between local short-term uses of man's environment and the maintenance and enhancement of long-term productivity, and

(v) Any irreversible and irretrievable commitments of resources which would be involved in the proposed action should it be implemented.

Wilderness Act

This act is the backbone of federal wilderness protection. The bill provides a legal description of wilderness and sets up a system for protecting specific wilderness areas. A piece of land may meet the legal definition of wilderness, but that does not entitle it to protection automatically. If Congress votes to declare the land a "wilderness area," however, it is given special protective status and added to the National Wilderness Preservation System. Lands within this system cannot be developed or degraded. When the act was first passed in 1964, it established 9.1 million acres of wilderness areas and called for federal agencies to examine their holdings for lands with potential for wilderness designation. In 1994 nearly 96-million acres have been designated as wilderness areas. Following is the complete text of the act.

Public Law 88-577
88th Congress, S. 4
September 3, 1964

An Act

To establish a National Wilderness Preservation System for the permanent good of the whole people, and for other purposes.

Be it enacted by the Senate and House of Representatives of the United States of America in Congress assembled.

SHORT TITLE

Section 1. This Act may be cited as the "Wilderness Act".

WILDERNESS SYSTEM ESTABLISHED STATEMENT OF POLICY

Sec. 2 (a) In order to assure that an increasing population, accompanied by expanding settlement and growing mechanization, does not occupy and modify all areas within the United States and its possessions, leaving no lands designated for preservation and protection in their natural condition, it is hereby declared to be the policy of the Congress to secure for the American people of present and future generations the benefits of an enduring resource of wilderness. For this purpose there is hereby established a National Wilderness Preservation System, to be composed of federally owned areas designated by Congress as "wilderness areas", and these shall be administered for the use and enjoyment as wilderness, and so as to provide for the protection of these areas, the preservation of their wilderness character, and for the gathering and dissemination of information regarding their use and enjoyment as wilderness; and no Federal lands shall be designated as "wilderness areas" except as provided for in this Act or by a subsequent Act.

(b) The inclusion of an area in the National Wilderness Preservation System notwithstanding, the area shall continue to be managed by the Department and agency having jurisdiction thereover immediately before its inclusion in the National Wilderness Preservation System as a separate unit nor shall any appropriations be available for additional personnel stated as being required solely for the purpose of managing or administering areas solely because they are included within the National Wilderness Preservation System.

DEFINITION OF WILDERNESS

(c) A wilderness, in contrast with those areas where man and his own works dominate the landscape, is hereby recognized as an area where the earth and its community of life are untrammeled by man, where man himself is a visitor who does not remain. An area of wilderness is further defined to mean in this Act an area of undeveloped Federal land retaining its primeval character and influence, without permanent improvements or human habitation, which is protected and managed so as to preserve its natural conditions and which (1) generally appears to have been affected primarily by the forces of nature, with the imprint of man's work substantially unnoticeable; (2) has outstanding opportunities for solitude or a primitive and unconfined type of recreation; (3) has at least five thousand acres of land and is of sufficient size as to make practicable its preservation and use in an unimpaired condition; and (4) may also contain ecological, geological, or other features of scientific, educational, scenic, or historical value.

NATIONAL WILDERNESS PRESERVATION SYSTEM— EXTENT OF SYSTEM

Sec. 3. (a) All areas within the national forests classified at least 30 days before the effective date of this Act by the Secretary of Agriculture or the Chief of the Forest Service as "wilderness", "wild", or "canoe" are hereby designated as wilderness areas. The Secretary of Agriculture shall—

    (1) Within one year after the effective date of this Act, file a map and legal description of each wilderness area with the Interior and Insular Affairs Committees of the United States Senate and the House of Representatives, and such descriptions shall have

the same force and effect as if included in this Act: *Provided, however,* That correction of clerical and typographical errors in such legal descriptions and maps may be made.

(2) Maintain, available to the public, records pertaining to said wilderness areas, including maps and legal descriptions, copies of regulations governing them, copies of public notices of, and reports submitted to Congress regarding pending additions, eliminations, or modifications. Maps, legal descriptions, and regulations pertaining to wilderness areas within their respective jurisdictions also shall be available to the public in the offices of regional foresters, national forest supervisors, and forest rangers.

(b) The Secretary of Agriculture shall, within ten years after the enactment of this Act, review, as to its suitability or nonsuitability for preservation as wilderness, each area in the national forests classified on the effective date of this Act by the Secretary of Agriculture or the Chief of the Forest Service as "primitive" and report his findings to the President. The President shall advise the United States Senate and House of Representatives of his recommendations with respect to the designation as "wilderness" or other reclassification of each area on which review has been completed, together with maps and a definition of boundaries. Such advice shall be given with respect to not less than one-third of all the areas now classified as "primitive" within three years after the enactment of this Act, not less than two-thirds within seven years after the enactment of this Act, and the remaining areas within ten years after the enactment of this Act. Each recommendation of the President for designation as "wilderness" shall become effective only if so provided by an Act of Congress. Areas classified as "primitive" on the effective date of this Act shall continue to be administered under the rules and regulations affecting such areas on the effective date of this Act until Congress has determined otherwise. Any such area may be increased in size by the President at the time he submits his recommendations to the Congress by not more than five thousand acres with no more than one thousand two hundred and eighty acres of such increase in any one compact unit; if it is proposed to increase the size of any such area by more than five thousand acres or by more than one thousand two hundred and eighty acres in any one compact unit the increase in size shall not become effective until acted upon by Congress. Nothing herein contained shall limit the President in proposing, as part of his recommendations to Congress, the alteration of existing boundaries of primitive areas or recommending the addition of any contiguous area of national forest lands predominantly of wilderness value. Notwithstanding any other provisions of this Act, the Secretary of

Agriculture may complete his review and delete such area as may be
necessary, but not to exceed seven thousand acres, from the southern
tip of the Gore Range-Eagles Nest Primitive Area, Colorado, if the
Secretary determines that such action is in the public interest.

(c) Within ten years after the effective date of this Act the Secretary
of the Interior shall review every roadless area of five thousand
contiguous acres or more in the national parks, monuments and
other units of the National Park System and every such area of, and
every roadless island within, the national wildlife refuges and game
ranges, under his jurisdiction on the effective date of this Act and
shall report to the President his recommendation as to the suitability
or nonsuitability of each such area or island for preservation as
wilderness. The President shall advise the President of the Senate
and the Speaker of the House of Representatives of his recommendation with respect to the designation as wilderness of each
such area or island on which review has been completed, together
with a map thereof and a definition of its boundaries. Such advice
shall be given with respect to not less than one-third of the areas and
islands to be reviewed under this subsection within three years after
enactment of this Act, not less than two-thirds within seven years of
enactment of this Act, and the remainder within ten years of enactment of this Act. A recommendation of the President for designation
as wilderness shall become effective only if so provided by an Act of
Congress. Nothing contained herein shall, by implication or otherwise,
be construed to lessen the present statutory authority of the Secretary
of the Interior with respect to the maintenance of roadless areas
within units of the national park system.

(d) (1) The Secretary of Agriculture and the Secretary of the Interior
shall, prior to submitting any recommendations to the President with
respect to the suitability of any area for preservation as wilderness—

  (A) give such public notice of the proposed action as they
      deem appropriate, including publication in the Federal
      Register and in a newspaper having general circulation
      in the area or areas in the vicinity of the land;

  (B) hold a public hearing or hearings at a location or locations
      convenient to the area affected. The hearings shall be an-
      nounced through such means as the respective Secretaries
      involved deem appropriate, including notices in the Fed-
      eral Register and in newspapers of general circulation in
      the area: *Provided,* That if the lands involved are located
      in more than one State, at least one hearing shall be held
      in each State in which a portion of the land lies;

(C) at least thirty days before the date of a hearing advise the Governor of each State and the governing board of each county, or in Alaska the borough, in which the lands are located, and Federal departments and agencies concerned, and invite such officials and Federal agencies to submit their views on the proposed action at the hearing or by no later than thirty days following the date of the hearing.

(2) Any views submitted to the appropriate Secretary under the provisions of (1) of this subsection with respect to any area shall be included with any recommendations to the President and to Congress with respect to such area.

(e) Any modification or adjustment of boundaries of any wilderness area shall be recommended by the appropriate Secretary after public notice of such proposal and public hearing or hearings as provided in subsection (d) of this section. The proposed modification or adjustment shall then be recommended with map and description thereof to the President. The President shall advise the United States Senate and the House of Representatives of his recommendations with respect to such modification or adjustment and such recommendations shall become effective only in the same manner as provided for in subsections (b) and (c) of this section.

USE OF WILDERNESS AREAS

Sec. 4 (a) The purposes of this Act are hereby declared to be within and supplemental to the purposes for which national forests and units of the national park and national wildlife refuge systems are established and administered and—

(1) Nothing in this Act shall be deemed to be in interference with the purpose for which national forests are established as set forth in the Act of June 4, 1897 (30 Stat. 11), and the Multiple-Use Sustained Yield Act of June 12, 1960 (74 Stat. 215).

(2) Nothing in this Act shall modify the restrictions and provisions of the Shipstead-Nolan Act (Public Law 539, Seventy-first Congress, July 10, 1930; 46 Stat. 1020), and the Thye-Blatnik Act (Public Law 733, Eightieth Congress, June 22, 1948; 62 Stat. 568), and the Humphrey-Thye-Blatnik-Andersen Act (Public Law 607, Eighty-fourth Congress, June 22, 1956; 70 Stat. 326), as applying to the Superior National Forest or the regulations of the Secretary of Agriculture.

(3) Nothing in this Act shall modify the statutory authority under which units of the national park system are created. Further,

the designation of any area of any park, monument, or other unit of the national park system as a wilderness area pursuant to this Act shall in no manner lower the standards evolved for the use and preservation of such park, monument, or other unit of the national park system in accordance with the Act of August 25, 1916, the statutory authority under which the area was created, or any other Act of Congress which might pertain to or affect such area, including, but not limited to, the Act of June 8, 1906 (34 Stat. 225; 16 U.S.C. 432 et seq.); section 3 (2) of the Federal Power Act (16 U.S.C. 796(2)); and the Act of August 21, 1935 (49 Stat. 666; 16 U.S.C. 461 et seq.).

(b) Except as otherwise provided in this Act, wilderness areas shall be devoted to the public purposes of recreational, scenic, scientific, educational, conservation, and historical use.

PROHIBITION OF CERTAIN USES

(c) Except as specifically provided for in this Act, and subject to existing private rights, there shall be no commercial enterprise and no permanent road within any wilderness area designated by this Act and, except as necessary to meet minimum requirements for the administration of the area for the purpose of this Act (including measures required in emergencies involving the health and safety of persons within the area), there shall be no temporary road, no use of motor vehicles, motorized equipment or motorboats, no landing of aircraft, no other form of mechanical transport, and no structure or installation within any such area.

SPECIAL PROVISIONS

(d) The following provisions are hereby made:

(1) Within wilderness areas designated by this Act the use of aircraft or motorboats, where these uses have already become established, may be permitted to continue subject to such restrictions as the Secretary of Agriculture deems desirable. In addition, such measures may be taken as may be necessary in the control of fire, insects and diseases, subject to such conditions as the Secretary deems desirable.

(2) Nothing in this Act shall prevent within national forest wilderness areas any activity, including prospecting, for the purpose of gathering information about mineral or other resources, if such activity is carried on in a manner compatible with the preservation of the wilderness environment. Furthermore, in accordance with such program as the Secretary of the Interior shall develop and conduct in

consultation with the Secretary of Agriculture, such areas shall be surveyed on a planned, recurring basis consistent with the concept of wilderness preservation by the Geological Survey and the Bureau of Mines to determine the mineral values, if any, that may be present; and the results of such surveys shall be made available to the public and submitted to the President and Congress.

(3) Notwithstanding any other provisions of this Act, until midnight December 31, 1983, the United States mining laws and all laws pertaining to mineral leasing shall, to the same extent as applicable prior to the effective date of this Act, extend to those national forest lands designated by this Act as "wilderness areas"; subject, however, to such reasonable regulations governing ingress and egress as may be prescribed by the Secretary of Agriculture consistent with the use of the land for mineral location and development and exploration, drilling, and production, and use of land for transmission lines, waterlines, telephone lines, or facilities necessary in exploring, drilling, producing, mining, and processing operations, including where essential the use of mechanized ground or air equipment and restoration as near as practicable of the surface of the land disturbed in performing prospecting, location, and, in oil and gas leasing, discovery work, exploration, drilling, and production, as soon as they have served their purpose. Mining locations lying within the boundaries of said wilderness areas shall be held and used solely for mining or processing operations and uses reasonably incident thereto; and hereafter, subject to valid existing rights, all patents issued under the mining laws of the United States affecting national forest lands designated by this Act as wilderness areas shall convey title to the mineral deposits within the claim, together with the right to cut and use so much of the mature timber therefrom as may be needed in the extraction, removal, and beneficiation of the mineral deposits, if needed timber is not otherwise reasonably available, and if the timber is cut under sound principles of forest management as defined by the national forest rules and regulations, but each such patent shall reserve to the United States all title in or to the surface of the lands and products thereof, and no use of the surface of the claim or the resources therefrom not reasonably required for carrying on mining or prospecting shall be allowed except as otherwise expressly provided in this Act: *Provided,* That, unless hereafter specifically authorized, no patent within wilderness areas designated by this Act shall issue after December 31, 1983,

except for the valid claims existing on or before December 31, 1983. Mining claims located after the effective date of this Act within the boundaries of wilderness areas designated by this Act shall create no rights in excess of those rights which may be patented under the provisions of this subsection. Mineral leases, permits, and licenses covering lands within national forest wilderness areas designated by this Act shall contain such reasonable stipulations as may be prescribed by the Secretary of Agriculture for the protection of the wilderness character of the land consistent with the use of the land for the purposes for which they are leased, permitted, or licensed. Subject to valid rights then existing, effective January 1, 1984, the minerals in lands designated by this Act as wilderness areas are withdrawn from all forms of appropriation under the mining laws and form disposition under all laws pertaining to mineral leasing and all amendments thereto.

(4) Within wilderness areas in the national forests designated by the Act, (1) the President may, within a specific area and in accordance with such regulations as he may deem desirable, authorize prospecting for water resources, the establishment and maintenance of reservoirs, water-conservation works, power projects, transmission lines, and other facilities needed in the public interest, including the road construction and maintenance essential to development and use thereof, upon his determination that such use or uses in the specific area will better serve the interests of the United States and the people thereof than will its denial, and (2) the grazing of livestock, where established prior to the effective date of this Act, shall be permitted to continue subject to such reasonable regulations as are deemed necessary by the Secretary of Agriculture.

(5) Other provisions of this Act to the contrary notwithstanding, the management of the Boundary Waters Canoe Apeak, formerly designated as the Superior, Little Indian Sioux, and Caribou Roadless Areas, in the Superior National Forest, Minnesota, shall be in accordance with regulations established by the Secretary of Agriculture in accordance with the general purpose of maintaining, without unnecessary restrictions on other uses, including that of timber, the primitive character of the area, particularly in the vicinity of lakes, streams, and portages: *Provided,* That nothing in this Act shall preclude the continuance within the area of any already established use of motorboats.

(6) Commercial services may be performed within the wilderness areas designated by this Act to the extent necessary for

activities which are proper for realizing the recreational or other wilderness purposes of the areas.

(7) Nothing in this Act shall constitute an express or implied claim or denial on the part of the Federal government as to exemption from State water laws.

(8) Nothing in this Act shall be construed as affecting the jurisdiction or responsibilities of the several States with respect to wildlife and fish in the national forests.

STATE AND PRIVATE LANDS WITHIN WILDERNESS AREAS

SEC 5. (a) In any case where State-owned or privately owned land is completely surrounded by national forest lands within areas designated by this Act as wilderness, such State or private owner shall be given such rights as may be necessary to assure adequate access to such State-owned or privately owned land by such State or private owner and their successors in interest, or the State-owned land or privately owned land shall be exchanged for federally-owned land in the same State of approximately equal value under authorities available to the Secretary of Agriculture: *Provided, however,* That the United States shall not transfer to a State or private owner any mineral interests unless the State or private owner relinquishes or causes to be relinquished to the United States the mineral interest in the surrounded land.

(b) In any case where valid mining claims or other valid occupancies are wholly within a designated national forest wilderness area, the Secretary of Agriculture shall, by reasonable regulations consistent with the preservation of the area as wilderness, permit ingress and egress to such surrounded areas by means which have been or are being customarily enjoyed with respect to other such areas similarly situated.

(c) Subject to the appropriation of funds by Congress, the Secretary of Agriculture is authorized to acquire privately owned land within the perimeter of any area designated by this Act as wilderness if (1) the owner concurs in such acquisition or (2) the acquisition is specifically authorized by Congress.

GIFTS, BEQUESTS, AND CONTRIBUTIONS

Sec. 6. (a) The Secretary of Agriculture may accept gifts or bequests of land within wilderness areas designated by this Act for Preservation as wilderness. The Secretary of Agriculture may also accept gifts or bequests of land adjacent to wilderness areas designated by this Act for preservation as wilderness if he has given sixty days advance notice thereof to the President of the Senate and the Speaker of the House of Representatives. Land accepted by the Secretary of Agriculture under this section shall become part of the wilderness area involved. Regulations with regard to any such land may be in accordance with such agreements, consistent with the policy of this Act, as are made at the time of such gift, or such conditions, consistent with such policy, as may be included in, and accepted with, such bequest.

(b) The Secretary of Agriculture or the Secretary of the Interior is authorized to accept private contributions and gifts to be used to further the purposes of this Act.

ANNUAL REPORTS

SEC. 7. At the opening of each session of Congress, the Secretaries of Agriculture and Interior shall jointly report to the President for transmission to Congress on the status of the wilderness system, including a list and descriptions of the areas in the system, regulations in effect, and other pertinent information, together with any recommendations they may care to make.

Approved September 3, 1964.

5

Protected Federal Lands

THIS CHAPTER PROVIDES BRIEF DESCRIPTIONS of all the national parks and national monuments, as well as listings of other segments of the park system that contain wilderness. Separate tables include all units in the National Wildlife Refuges and the National Wilderness Preservation System.

National Parks

All national parks are made up primarily of wilderness. As national parks they are protected to a large degree. Development, such as building roads, facilities, or dams, is at the discretion of the NPS. Development is only prohibited by law in areas that are designated as wilderness areas by Congress. In the listings below, the total acres of each unit are included, as well those acres that are wilderness areas, if any have been so designated.

Acadia National Park, Maine
Route 1, Box 177
Bar Harbor, ME 04609
(207) 288-9561
Established: 1919
Total acres: 41,409

Acadia has the distinction of being the only national park in the northeastern United States. It was originally established as Sieur de Monts

National Monument in 1916, changed to Lafayette National Park in 1919, and finally Acadia National Park in 1929.

This park is known for its rocky coastline, with pine covered, craggy cliffs falling into the sea. Acadia also has many inland lakes, and during the course of the year is inhabited by more than 275 species of birds.

Logging operations were common in the area before Acadia became a park, and several millionaires owned summer homes there. Most of the land that makes up the park today was donated; much of it by John D. Rockefeller, Jr.

Arches National Park, Utah
P.O. Box 907
Moab, UT 84532
(801) 259-8161
Established: 1971
Total acres: 73,379

In the dry desert valleys of eastern Utah, just west of the Rocky Mountains, sits Arches National Park. This is one park that is aptly named, for more than 90 stone arches carved by natural elements have been found within the park boundaries. The area also contains many other geological formations such as pinnacles and spires made of sandstone.

Arches was originally proclaimed as a national monument in 1929, and then upgraded to a national park in 1971. Nearby Canyonlands National Park is considered a sister park to Arches and has many similar formations.

Badlands National Park, South Dakota
Box 6
Interior, SD 57750
(605) 433-5361
Established: 1978
Total acres: 243,244
Wilderness area acres: 64,250

More than 60 million years ago the area now surrounding the White River in South Dakota was a shallow inland sea. When the Rocky Mountains and Black Hills pushed up through the earth's crust, the sea drained off, leaving a plain of sediment and mud several hundred feet deep. Today that sediment and mud has eroded away to form constantly changing craggy spires and pinnacles in what is now Badlands National Park.

Badlands is in the heart of traditional Sioux and Arikara Indian territory. It was originally authorized as a national monument in 1929. The park is considered to be an important scientific region due to a large number of prehistoric animal fossils that have been discovered there.

Big Bend National Park, Texas
TX 79834
(915) 477-2251
Authorized: 1935
Established: 1944
Total acres: 802,541

Located on the border between Mexico and Texas, Big Bend is one of the most isolated national parks in the contiguous United States. El Paso, at 300 miles away, is the nearest major city. The closest towns are more than 70 miles away. This makes Big Bend one of the more peaceful parks in the system, with less than 500,000 visitors annually.

Big Bend is situated in the desert along a big bend in the Rio Grande River. Millions of years ago the area was first an inland sea, and then later it became tropical forests and marshes, complete with giant crocodiles and dinosaurs. The fossilized bones of the largest flying creature ever known were found within the park in 1975. Today the Rio Grande runs through spectacular canyons carved into the desert rocks and mountains.

Biscayne National Park, Florida
Box 1369
Homestead, FL 33030
(305) 247-2044
Established: 1980
Total acres: 181,500

At the southeastern tip of Florida lies Biscayne Bay and Biscayne National Park, which encompasses the southern half of the bay and about 30 small islands, including Elliot Key. The park protects marine life, rare tropical plants, and migratory birds. Nearby is Key Largo Coral Reef Marine Sanctuary, which covers 100 square miles and includes the only living coral reefs within the contiguous United States.

Bryce Canyon National Park, Utah
Bryce Canyon, UT 84717
(801) 834-5322
Established: 1924
Total acres: 37,102

The rock formations found in Bryce Canyon National Park are unlike any other in the world. The entire canyon is filled with what look like giant pink, orange, and red sand castles, many towering hundreds of feet high. Many have names, such as Hindu Temples, Gulliver's Castle, and Thor's Hammer.

Named for an early settler and rancher, Bryce was originally classified as a national monument by presidential decree in 1923. The following year it was designated as Utah National Park, though the name was changed back to Bryce a few years later.

Canyonlands National Park, Utah
446 South Main Street
Moab, UT 84532
(801) 259-7164
Established: 1964
Total acres: 337,570

At the confluence of the Green and Colorado rivers in southeast Utah sits Canyonlands National Park. The park is known for its striking rock formations, including arches, spires, flat-topped mesas, and canyons. Just south of the park is Lake Powell, which feeds the Grand Canyon. Thirty miles to the northeast is Arches National Park.

Capitol Reef National Park, Utah
Torrey, UT 84775
(801) 425-3791
Established: 1971
Total acres: 241,865

Prospectors nicknamed this area "reef" because it blocked passage to covered wagons, like a coral reef blocks the passage of a ship. Other explorers applied the name capitol because some of the rock formations appeared dome-like. The park is located in south-central Utah. The Capitol Reef geological formations are part of what is called the Waterpocket Fold: a 100-mile-long fold in the earth's crust. Seventy-two miles of the fold are within the park, but only 20 miles make up what is actually known as the Capitol Reef.

Carlsbad Caverns National Park, New Mexico
3225 National Parks Highway
Carlsbad, NM 88220
(505) 885-8884
Established: 1930
Total acres: 46,753
Wilderness area acres: 33,125

More than 70 different caves can be found within the boundaries of Carlsbad Caverns National Park. The largest and best known of these is Carlsbad Cavern, which is open to the public through guided tours. This particular cavern has many rooms of interest, with magnificent stalactites and stalagmites. The most striking is The Big Room, which is considered the largest underground chamber in the world with a ceiling 225 feet high and a floor more than 4,000 square feet. This park was first protected as a national monument in 1923, before being made into a national park seven years later.

Channel Islands National Park, California
1901 Spinnaker Drive
Ventura, CA 93001
(805) 644-8157
Established: 1980
Total acres: 249,354

Off the coast of California lie eight Channel Islands. San Clemente and San Nicolas Islands are owned by the Navy and used for bombing practice. Santa Catalina is privately owned and developed as a resort. Santa Cruz Island is mostly owned and protected by The Nature Conservancy. Santa Rosa is also privately owned, and is used as a cattle ranch.

Santa Barbara and Anacapa Islands were made a national monument in 1938. San Miguel was added to the monument in 1976, and today these three make up Channel Islands National Park. The islands are important refuges for several species of birds and seals, and are considered to have the best scuba diving in southern California.

Crater Lake National Park, Oregon
Box 7
Crater Lake, OR 97604
(503) 594-2211
Established: 1902
Total acres: 183,224

At nearly 2,000 feet deep, Crater Lake is the deepest in the United States. The lake was formed after mount Mazama in Oregon's Cascade range erupted about 6,840 years ago, forming a crater basin 20 square miles in area. The national park includes the lake itself as well as surrounding peaks, meadows, and pine forests.

Death Valley National Park, California–Nevada
Death Valley, CA 92328
(619) 786-2331
Authorized: 1994
Total acres: 3,336,528
Wilderness area acres: 3,158,038

For pioneers heading to the West coast in the 1800s, Death Valley stood as one last obstacle after an already long and difficult trip. The lowest point in the Western Hemisphere is found here, at 282 feet below sea level, as well as Telescope Peak, at 11,049 feet. Death Valley was originally established as a national monument in 1933. In 1994 the California Desert Protection Act authorized the unit to become a national park and designated most of it as a wilderness area.

Denali National Park, Alaska
Box 9
Denali National Park, AK 99755
(907) 683-2294
Established: 1917
Total acres: 4,716,726
Wilderness area acres: 2,124,783

Originally called Mount McKinley National Park before the name was changed in 1980, Denali is best known as home to the highest mountain peak in the United States, the 20,320-foot Mount McKinley. Early natives actually called the mountain, *Denali,* or "the high one." Today the national park is joined with a national preserve of 1.33 million acres to make a national park and preserve of over 6 million acres.

The lowest point within Denali is 1,400 feet high, and in addition to McKinley, the park contains several other impressive mountain peaks, including the 17,400-foot Mount Foraker, known as "Denali's Wife," and the 13,220-foot Mount Silverthrone.

As with other wilderness regions in Alaska, much of Denali is quite isolated from human impact, leaving the natural ecology for the most part undisturbed. Several varieties of spruce, aspen, birch, cottonwood, and willow trees grow in the lower elevations of the park, up to the timberline at about 3,000 feet. Above the timberline the only vegetation is in the form of mosses, lichens, grasses, and hearty plants capable of surviving in the frigid, tundra-like environment. Animal species include caribou, moose, wolves, foxes, Dall sheep, golden and bald eagles, and grizzly bears.

Dry Tortugas National Park, Florida
c/o Everglades National Park
P.O. Box 279
Homestead, FL 33030
(305) 247-6211
Established: 1992
Total acres: 64,700

The Dry Tortugas is a group of small islands, or keys, located 70 miles west of Key West, off the coast of Florida. A lack of fresh water and an abundance of turtles *(tortugas* in Spanish) led to the name. Pirates frequented the area until 1821 when the United States government took control. In 1846 the government began building Fort Jefferson, which covers nearly all of Garden Key, to protect shipping lanes in the Gulf of Mexico. The fort was later turned into a prison, and then abandoned after an epidemic of yellow fever in 1874. The area was first made Fort Jefferson National Monument before becoming a national park.

Coral reefs surrounding the keys are famous for turtles, as well as yellowtail, grouper, snapper, sharks, barracuda, lobsters, and shells.

Everglades National Park, Florida
P.O. Box 279
Homestead, FL 33030
(305) 247-6211
Established: 1934
Total acres: 1,400,533
Wilderness area acres: 1,296,500

Everglades National Park protects the southern section of the Everglades, which originally covered 4-million acres. Today the Everglades is less than half its original size, but is still the largest fresh water marsh in the world.

The grassy marsh itself actually acts as one gigantic, slowly moving river. Water enters from the Lake Okeechobee in the north and moves south across the tip of Florida to finally enter the Atlantic Ocean through Florida Bay. The ecosystem is home to birds of many species, including egrets, herons, roseate spoonbills, brown pelicans, and wood ibis (the only storks in the United States). Crocodiles may be the most famous residents of the Everglades, but the park is also home to deer, raccoons, possums, cougars, and black bears.

Modern development outside the park boundaries has reduced the amount of fresh, clean water that enters the Everglades, and is threatening the very existence of the marsh. In 1993 President Clinton and Secretary of the Interior Bruce Babbitt unveiled an extensive plan to restore the region.

Gates of the Arctic National Park, Alaska
Box 74680
Fairbanks, AK 99707
(907) 456-0281
Established: 1980
Total acres: 7,952,000
Wilderness area acres: 7,167,192

Gates of the Arctic National Park is four times the size of Yellowstone, but almost completely isolated. No roads provide access to the park, which is located 200 miles northwest of Fairbanks. Air service does land 20 miles south of the park, at Bettles, and from there visitors must either walk or hire a bush pilot who can land on lakes and rivers to drop them off in the park.

The most striking feature in Gates of the Arctic are the peaks of the Brooks Range, as well as the valleys and glaciers that are still carving them. In the 1930s Robert Marshall named two facing peaks Frigid

Crags and Boreal, and together called them the Gates of the Arctic. He was the first to push for wilderness protection for the area.

Glacier Bay National Park and Preserve, Alaska
Gustavus, AK 99826
(907) 697-2232
Established: 1980
Total acres: 3,225,284
Wilderness area acres: 2,664,840

Glacier Bay is about 50 miles long and varies between 2.5 and 10 miles wide. It is named for the glaciers which spill down from valleys and drop abruptly into the sea, with walls of ice up to 200 feet high. The glaciers are receding at a phenomenal rate. Only 250 years ago, the entire bay was covered with ice 3,000 feet thick. The first explorer who was actually able to sail into the bay was John Muir in 1879.

In 1925 Glacier Bay was first set aside as a national monument. The protected area was expanded when it was made a national park 55 years later, and today the park encompass both the bay itself and the surrounding area, including an area of rain forest, which was also covered with ice only a few hundred years ago.

Glacier National Park, Montana
West Glacier, MT 59936
(406) 888-5441
Established: 1910
Total acres: 1,013,598

Located along the continental divide in northern Montana, Glacier National Park is connected to Canada's Waterton National Park. Together these two units make up Glacier–Waterton International Peace Park. This park is known for its many glaciers and glacier-carved granite peaks and valleys, as well as abundant wildlife, including grizzly bears, bighorn sheep, moose, mountain goats, elk, and deer. The area is also home to more than 200 species of birds, from hawks and eagles to thrushes and wren.

Grand Canyon National Park, Arizona
Box 129
Grand Canyon, AZ 86023
(602) 638-7888
Established: 1919
Total acres: 1,218,375

President Theodore Roosevelt called the Grand Canyon " the one great sight which every American should see." Many visitors from the United States and around the world agree, making Grand Canyon National Park one of the most popular in the country, with close to 2 million visitors annually. The canyon itself varies in width from one-tenth of a mile to 18 miles across, and is more than a mile deep—5,800 feet at its deepest point.

Geologists differ in opinion on exactly how the canyon was formed, but agree that it was most likely carved in some way by what is now the Colorado River. The park was the source of controversy in the 1960s when plans surfaced to build a dam that would flood parts of the canyon. This plan was defeated by the lobbying efforts of a coalition of environmental groups (see Chapter 2, p. 43–44). Today the canyon lies between two large reservoirs, Lake Mead formed by Hoover Dam to the west, and Lake Powell behind Glen Canyon Dam to the east.

Grand Teton National Park, Wyoming
P.O. Drawer 170
Moose, WY 83012
(307) 733-2880
Established: 1929
Total acres: 310,418

At 13,770, the Grand Teton is the highest in a 40-mile-long chain of 11 major peaks known collectively as The Tetons. These mountains are especially impressive to the eye because they rise straight up off a flat plain, with no foothills. South of the range is Jackson Hole, home of forests, meadows, abundant wildlife, and Jackson Lake, and bisected by the Snake River. John D. Rockefeller was largely responsible for the creation of this park when he purchased 30,000 acres of the Jackson Hole Basin and presented it to the U.S. government for preservation.

Great Basin National Park, Nevada
Baker, NV 89311
(702) 234-7331
Established: 1986
Total acres: 77,109

Just inside Nevada, along the Utah border, lies the South Snake Range and Great Basin National Park. The Great Basin itself is a valley carved by ancient glaciers. The park contains the only remaining glacier in Nevada today, as well as landscapes varying from dry desert, to forests, to the 13,063-foot Wheeler Peak.

Great Smoky Mountains National Park, Tennessee
Gatlinburg, TN 37738
(615) 436-1201
Authorized: 1926
Established: 1940
Total acres: 520,269

The Smoky Mountains are a part of the larger Appalachian mountain chain. These are among the oldest mountains in the world, first formed when they were on the bottom of a shallow sea around 500 million years ago. When the Interior Department decided to create national parks in the East during the 1920s, the Great Smoky Mountains, Shenandoah Valley, and Mammoth Cave were all chosen as sites, but the land was privately owned, mostly by logging companies. Half of the necessary money to purchase the Great Smoky Mountains, $5 million, was raised through contributions and appropriations. The other half was donated by John D. Rockefeller, and the park was dedicated by Franklin D. Roosevelt in 1940.

Great Smoky Mountains National Park follows the ridgeline, which is also the North Carolina–Tennessee border, for 72 miles over pine-covered peaks up to 6,000 feet high. The park is located in both states.

Guadalupe Mountains National Park, Texas
H.C. 60, Box 400
Salt Flat, Texas 79847
(915) 828-3251
Authorized: 1966
Established: 1972
Total acres: 86,416
Wilderness area acres: 46,850

The Guadalupe Mountains stand like islands in the vast, level desert of northern Texas. Guadalupe Peak is the highest point in the state at 8,749 feet. Most of the land that now makes up the park, 72,000 acres, was procured privately by a judge from the area, J. C. Hunter, to be made a park. After he died his son sold the land to the government for $21 per acre, and another rancher, Wallace E. Pratt, donated an additional 5,632 acres.

Guadalupe National Park contains several different types of environments. The lowlands are a dry, desert region, with cactus and agave plants, while as the elevation increases the environment gradually changes to an alpine, forested region.

Haleakala National Park, Hawaii
Box 369
Makawao, HI 96768
(808) 572-9306
Established: 1960
Total acres: 28,655
Wilderness area acres: 19,270

Haleakala is a dormant volcano located on the island of Maui in Hawaii. No eruptions have occurred in the area since the 1700s, and today

Haleakala is a popular spot for visitors to the island. Most of Haleakala National Park is located within the crater itself, which is 7.5 miles long by 2.5 miles wide and resembles what one might expect the surface of Mars to look like. Smaller volcanic cinder cones stick up sporadically through the rocky red and orange floor.

Little vegetation survives in the pumice and ash soil, though one plant, the yucca-like silversword, is found nowhere else on earth but Hawaii's volcanic terrain. This plant lives between 7 and 20 years before flowering brilliantly for the first time, and then dies once it has dropped its seeds.

Accessible by automobile, the summit of Haleakala is 10,023 feet, and on a clear day this spot affords views across the sea to the "Big Island," Lanai, Molokai, and Oahu.

Hawaii Volcanoes National Park, Hawaii
HI 96718
(808) 967-7311
Established: 1916
Total acres: 229,117
Wilderness area acres: 123,100

Hawaii Volcanoes National Park includes two active volcanoes on Hawaii's Big Island, 13,680-foot Mauna Loa and 4,090-foot Kilauea. Both of these volcanoes sporadically erupt in the present day, and watching their lava flows is a popular spectator sport on the island, though the lava occasionally destroys roads and houses. Generally the flows are slow and predictable enough to not pose a danger to onlookers.

Hot Springs National Park, Arkansas
P.O. Box 1860
Hot Springs, AR 71902
(501) 624-3383
Established: 1921
Total acres: 5,839

Hot Springs National Park is located along a fault, or crack in the earth's crust and contains 47 natural thermal springs, from which flows a million gallons of water per day at an average temperature of 143 degrees. Concessionaires operate bathhouses for visitors in various spots throughout the park.

Isle Royale National Park, Michigan
87 N. Ripley St.
Houghton, MI 49931
(906) 482-0986
Established: 1931

Total acres: 571,790
Wilderness area acres: 132,018

At 45 miles long and 9 miles wide, Isle Royale is the largest island in Lake Superior. It is located 18 miles from the Minnesota shore, 45 miles from the closest Michigan shore, and 15 miles from the shore of Canada. The soil varies from only a few inches to a few feet deep, but the island is forested and hosts a wide variety of animal species, including beaver, mink muskrat, weasel, and bald eagles.

Moose did not make their way onto Isle Royale until the early twentieth century, and are thought to have crossed over from the mainland when Superior froze in 1912. Without any predators their numbers skyrocketed, but during the winter of 1948–1949 wolves made their way across and now help keep the moose population in check.

Joshua Tree National Park, California
74485 National Monument Drive
Twentynine Palms, CA 92277
(619) 367-7511
Authorized: 1994
Total acres: 793,950
Wilderness area acres: 561,470

Rocky hills, valleys, cliffs, and outcroppings of giant boulders dot the desert here in southern California. Surrounding the rocks are a variety of up to 20-foot-high yucca plants named Joshua Trees by Mormon settlers. The area is a popular spot for rock climbers to test their skills. It was originally a national monument but was authorized to become a national park by Congress in 1994.

Katmai National Park and Preserve, Alaska
P.O. Box 7
King Salmon, AK 99613
(907) 246-3305
Established: 1980
Total acres: 3,960,000
Wilderness area acres: 3,384,385

In 1912 a series of mighty, violent volcanic eruptions spewed forth pumice and lava from a hundred new fissures in southern Alaska. Several new volcanoes were formed, clouds of dust filled the sky, and lava covered an area of 42 square miles up to 300 feet deep. In 1916 one valley, named the Valley of Ten Thousand Smokes, was found to still have thousands of fumaroles spewing acidic steam up to 1,000 feet into the air. By 1918 President Woodrow Wilson set aside the area as a national monument and named it after one of the volcanoes, Mount Katmai.

Today less than a dozen fumaroles remain and Mount Katmai is filled with a beautiful, jade-colored lake. Among other animals, the park is home to moose, wolves, minx, lynx, and bear, including the world's largest land carnivore at 600 to 1,300 pounds, the Alaskan brown bear.

Kenai Fjords National Park, Alaska
Box 1727
Seward, AK 99664
(907) 224-3874
Established: 1980
Total acres: 570,000

On the Kenai Peninsula in southern Alaska, retreating ice and glaciers carved steep valleys, or fjords, dropping dramatically into the sea. The 720-square-mile Harding Ice Field still remains, and is within the park's boundaries.

Kings Canyon National Park, California
Three Rivers, CA 93271
(209) 565-3341
Established: 1940
Total acres: 461,901
Wilderness area acres: 456,552

Kings Canyon National Park is connected to and managed together with Sequoia National Park in the Sierra Nevada Mountains. Kings Canyon offers some of the finest areas for back country trips in the United States. Most of this wilderness is only open to a limited number of backpackers each year, and contains numerous lakes, forests, and jagged peaks. Wildlife includes black bears, deer, marmots, and foxes, as well as rainbow and brook trout in the lakes and streams.

Kobuk Valley National Park, Alaska
P.O. Box 1029
Kotzebue, AK 99752
(907) 442-3890
Established: 1980
Total acres: 1,750,421
Wilderness area acres: 190,000

At 40 miles north of the Arctic Circle in northwestern Alaska, Kobuk Valley National Park is one of the most isolated parks in the state. This area also receives less attention than some of the state's more celebrated parks, such as Denali or Glacier Bay, but it is a beautiful region in its own right. The Kobuk River winds through both forests and ancient tundra. The Great Kobuk Sand Dunes rise up to 100 feet and spread over 25 square miles.

At Onion Portage, caribou have crossed the Kobuk River during their migrations for more than 10,000 years. Eskimos have camped there to hunt them for nearly as long, and scientists consider this an important archeological site.

Lake Clark National Park, Alaska
701 C Street
P.O. Box 61
Anchorage, AK 99513
(907) 271-3751
Established: 1980
Total acres: 3,000,000
Wilderness area acres: 2,619,550

Lake Clark National Park, in southern Alaska, is only 150 miles from Anchorage, making it one of the more accessible parks in the state, even though no roads go there. Small water planes ferry visitors from Anchorage, Kenai, or Lliamna, a village 35 miles west of the park.

This park is full of lakes of all sizes, surrounded by towering mountains. Lake Clark itself is the largest lake at 40 miles long. The shoreline of Cook Inlet forms the park's southeastern boundary, and with direct access to the sea, several lakes in and around the park are important salmon spawning grounds.

Lassen Volcanic National Park, California
P.O. Box 100
Mineral, CA 96063
(916) 595-4444
Established: 1916
Total acres: 106,372
Wilderness area acres: 78,982

Mount Lassen Volcano remained dormant for around 400 years before a seven-year series of eruptions began in 1914. It was during this active period that Lassen Peak and the surrounding area was made a national park. Today the volcano is once again dormant, though future eruptions are possible.

Mount Lassen is the southernmost in a chain of volcanoes known as the Cascades, stretching up through the Pacific Northwest from California into Canada. The Cascade Range includes such other volcanoes as Mount Shasta, Mount Hood, and Mount Rainier. The last in the chain to erupt was Mount St. Helens in 1980.

Mammoth Cave National Park, Kentucky
Mammoth Cave, KY 42259
(502) 758-2251

Authorized: 1926
Established: 1941
Total acres: 52,419

Although not yet fully mapped, Mammoth Cave in southern Kentucky contains almost 200 miles of charted passageways. These caves, discovered sometime in the 1790s, have a considerable history. During the war of 1812 nitrate from the cave's dirt was used to make gunpowder. Later in that century the cave became popular for performing artists, as a Shakespearean actor, a Norwegian violinist, and a Swedish vocalist all performed inside the caverns. During the 1840s a doctor's theory about the constant temperature and clean air turned part of the cave into a sanitarium where patients lived in stone huts. Today the cave is open to tours guided by the NPS.

Mesa Verde National Park, Colorado
Mesa Verde National Park, CO 81330
(303) 529-4465
Established: 1906
Total acres: 52,085
Wilderness area acres: 8,100

American Indians lived and flourished in the area now comprising Mesa Verde National Park for 800 years before abandoning the area for reasons unknown sometime in the thirteenth century. What is left behind are spectacular stone villages built beneath massive overhanging cliffs.

The natives, known now as the Anasazi, originally lived along streams and on top of mesas, but the buildings built later and protected by the cliff walls are the ones that remain intact today. When Mesa Verde was originally made a park in 1906, most of the more prominent ruins were not included because they were on part of the Ute Indian Reservation. In 1913 the Utes exchanged these areas for a larger acreage of land in a different location.

Mojave National Preserve, California
74485 National Monument Drive
Twentynine Palms, CA 92277
(619) 367-7511
Authorized: 1994
Total acres: 1,458,000
Wilderness area acres: 695,216

The California Desert Protection Act of 1994 authorized the creation of this newest national preserve in southeastern California. It includes portions of the Sonoran Desert, high elevation Mojave Desert, and Great Basin Desert. Within preserve boundaries are 16 mountain ranges with peaks up to 6,000 feet, cinder cones, and extensive sand dunes. Originally the area was to be a national park; the designation was changed to preserve shortly after the first printing of this book.

Mount Rainier National Park, Washington

Tahoma Woods, Star Route
Ashford, WA 98304
(206) 569-2211
Established: 1899
Total acres: 235,404
Wilderness area acres: 216,855

At 14,410 feet, Mount Rainier is the highest peak in the Cascade Range. This volcano last erupted around 2,000 years ago and is dormant today, though it could erupt again in the future. Steam vents found on the mountain attest to the existing pressure underneath. The peak is well known for its 41 glaciers, including the longest and largest glaciers in the lower 48 states. Though the mountain itself is mostly rocky and barren, Mount Rainier National Park is also known for its lush and beautiful pine forests below the 6,500-foot timberline.

North Cascades National Park, Washington

2105 Highway 20
Sedro Wooley, WA 98284
(206) 855-1331
Established: 1968
Total acres: 504,780
Wilderness area acres: 634,614

Along the Canadian border in northern Washington is North Cascades National Park. This park is known for its jagged peaks, numerous lakes, and glaciers. A dominant feature of the area is Ross Lake, which is actually one of three reservoirs formed by dams on the Skagit River. Ross Lake extends 1.5 miles into Ernest C. Manning Provincial Park on the Canadian side of the border.

Olympic National Park, Washington

600 East Park Avenue
Port Angeles, WA 98362
(206) 452-4501
Established: 1938
Total acres: 921,942
Wilderness area acres: 876,669

The only temperate rain forests in the United States are located in Olympic National Park, Washington. The forests on the westward side of 7,965-foot Mount Olympus receive 12 feet or more of rain (or melted snow) every year. Trees up to 300 feet high drip with damp moss and ferns cover the ground. The higher elevations include 60 glaciers within the park. The area first became an interest to preservationists because it is home to the Roosevelt elk, which can grow to 1,000 pounds. Surrounding the park today are national forests with numerous sections leveled by clear-cut logging. Olympic National Park remains a preserve of uncut, virgin forest.

Petrified Forest National Park, Arizona
Petrified Forest National Park, AZ 86028
(602) 524-6228
Established: 1962
Total acres: 93,532
Wilderness area acres: 50,260

Two hundred million years ago in what is now arid Arizona desert grew a lush, swampy forest. Many logs from these forests became buried by mud, sand, and volcanic ash. Theoretically, over time minerals filled in the wood cells and the logs virtually turned to stone. Today petrified logs, many of them shattered into fragments, cover the ground in certain areas of Petrified Forest National Park.

The call to protect the Petrified Forest, as it is known, began in the late 1800s when many of the logs were being destroyed. Many were dynamited in search of gems, others were shipped off as souvenirs, and a mill was actually built to crush the logs in order to use the minerals as abrasives. It was not until 1906, however, that the area was protected as one of the first national monuments created in this country with the passage of the Antiquities Act.

Redwood National Park, California
1111 Second Street
Crescent City, CA 95531
(707) 464-6101
Established: 1968
Total acres: 110,132

Coastal redwoods, found only in northern California and southern Oregon, are the tallest trees in the world. Many live to be over 1,000 years old. The tallest redwood recorded to date was measured at 368 feet high.

In 1963–1964, the NPS conducted a survey and determined that of an original redwood forest growth of nearly 2 million acres, only 15 percent remained, with 2.5 percent protected in state parks. With the rapid rate of logging in the area, it was estimated that nearly all of the remaining unprotected redwoods would be gone within 20 or 30 years.

When Redwood National Park was created in 1968, continued logging around the park's borders eroded an important watershed and sent silt into the streams, raising the water level and threatening to drown many protected groves. In 1976 the park was expanded to include formerly logged areas, and a major reforestation project was initiated.

Rocky Mountain National Park, Colorado
Estes Park, CO 80517
(303) 586-2371
Established: 1915
Total acres: 266,957

Rocky Mountain National Park in north-central Colorado contains 59 peaks over 12,000 feet, with Long's Peak the highest at 14,255. During the 1880s a mining boom brought settlers to the area temporarily, but when the mining did not pan out, most moved on. The push to preserve the area as a park was spearheaded by naturalist Enos Mills and the Colorado state legislature. Today the park protects forests, lakes, glaciers, and abundant wildlife, including the brown bear, bighorn sheep, elk, deer, and golden eagle.

Samoa National Park, American Samoa
c/o American Samoa Office of Tourism
P.O. Box 1147
Pago Pago
American Samoa 96799
(684) 699-9280
Established: 1992
Total acres: 8,870

The land that makes up Samoa National Park is not owned by the federal government, but leased from Samoan tribal chiefs. The park consists of three units on three separate islands. Tutuila contains a 2,800-acre rain forest, Ta'u includes a 5,400-acre rain forest, and Ofu, the smallest unit, is made up primarily of reef and beach. All three units include portions of reef.

Sequoia National Park, California
Three Rivers, CA 93271
(209) 565-3341
Established: 1890
Total acres: 846,989
Wilderness area acres: 280,428

This park is located side by side with Kings Canyon National Park in California's Sierra Nevada Mountains. Sequoia was the first to be established and is best known for its groves of giant sequoia. These trees are the largest living things on earth and can only be found here. The largest of these is the General Sherman Tree, which is 275 feet high and 103 feet around at the base.

Mount Whitney is also located within Sequoia National Park. This is the tallest spot in the continental United States, rising to 14,495 feet.

Shenandoah National Park, Virginia
Route 4, Box 348
Luray, VA 22835
(703) 999-2243
Established: 1926
Total acres: 195,382
Wilderness area acres: 79,579

When Shenandoah National Park was established it broke new ground in the way that national parks were conceived. Shenandoah was one of the first parks located on the East Coast and surrounded by large population centers. It was established at the same time as Great Smoky Mountains and Mammoth Caves national parks, and the only older park east of the Mississippi is Acadia. Unlike western parks, Shenandoah was not originally public land, but was owned by many private land holders. The land had to be acquired slowly through donations and both federal and state government appropriations.

Washington, D.C., is only 75 miles away from this national park, and Baltimore, Philadelphia, and New York are just a short drive away. Shenandoah provides a quiet refuge for people from both the cities and rural areas to enjoy a wilderness setting. The park itself is a thin band following the Blue Ridge Mountains and includes hiking trails and the famous Skyline Drive, which provides breathtaking views of the Shenandoah Valley to the west and Virginia Valley to the east.

Theodore Roosevelt National Park, North Dakota
P.O. Box 7
Medora, ND 58645
(701) 623-4466
Established: 1978
Total acres: 70,416
Wilderness area acres: 29,920

Known as the "Badlands of North Dakota," Theodore Roosevelt National Park is a harsh setting, with frigid winters and arid, infertile soil. In the 1880s Theodore Roosevelt visited the area on a hunting trip and decided to stay to start a cattle ranch. Roosevelt loved the outdoors, and this spot in particular, and would have been happy with his new vocation had it not lost money, but the cold winters killed off many cattle and forced him and other local ranchers out of business.

After Roosevelt died, in 1919, it was proposed to set aside some of this region as a memorial. In 1947 Congress established the Theodore Roosevelt National Memorial Park. Later it was upgraded to national park status. The area is known for sweeping valleys and strange, colorful erosional formations carved by the Little Missouri River.

Virgin Islands National Park, U.S. Virgin Islands

P.O. Box 7789
Charlotte Amalie
St. Thomas, VI 00801
(809) 775-6238
Established: 1956
Total acres: 14,689

Virgin Islands National Park encompasses about three-fourths of St. John Island and 5,000 acres under water. The park was designated to protect the island from excessive development. It includes tropical rainforests, volcanic peaks, unspoiled beaches, the ruins of Danish sugar plantations, and spectacular coral reefs.

Voyageurs National Park, Minnesota

P.O. Box 50
International Falls, MN 56649
(218) 283-9821
Authorized: 1971
Total acres: 218,036

On Rainy Lake, along the Canadian border in northern Minnesota, lies Voyageurs National Park. This is an isolated land of forests, lakes, and rivers and was once the domain of French-Canadian fur trappers and traders, or *Voyageurs*. Today the park is well known for hiking, canoe trips, and fishing, with 87,500 acres of water. Abundant wildlife in the region includes beaver, deer, black bear, moose, and timber wolves.

Wind Cave National Park, South Dakota

Hot Springs, SD 57747
(605) 745-4600
Established: 1903
Total acres: 28,292

As with any large cave, when the atmospheric pressure outside changes, air is forced either in or out of the mouth of the cave to equalize the pressure inside. It was this wind that led to the name of the cave, and eventually the park itself. This was the first cave to be added to the park system. It includes nearly 40 miles of explored tunnels and chambers, many with beautiful crystal formations on the walls. Outside the cave, the park protects prairie grasslands to the southeast of the Black Hills.

Wrangell–St. Elias National Park and Preserve, Alaska

P.O. Box 29
Glenallen, AK 99588
(907) 822-5235

Established: 1980
Park acres: 8,331,604
Preserve acres: 4,856,721
Wilderness area acres: 9,078,675

The largest park in the National Park System is Wrangell–St. Elias. The area under protection is even larger, as the park adjoins both Wrangell–St. Elias National Preserve and Kluane National Park in Canada. Seven peaks within Wrangell–St. Elias rise above 14,500 feet, including Mount St. Elias, which is the second highest peak in the United States at 18,008 feet. More than 100 major glaciers constitute the largest glacier system in the United States. The forested lowlands are home to abundant Dall sheep and grizzly bear. The Chitna River and Valley run down the center of the park and are home to two old mining towns, McCarthy and Kennicott, which have only a handful of residents today.

Yellowstone National Park, Wyoming–Montana–Idaho
P.O. Box 168
Yellowstone National Park, WY 82190
(307) 344-7381
Established: 1872
Total acres: 2,221,773

Yellowstone was the first national park in the world, and after more than 100 years, is still one of the crown jewels of our National Park System. When the park was first established in 1872, no system existed for its management and protection. An unpaid park supervisor was appointed, but without funds or man power he was unable to stop illegal logging and wildlife poaching. In 1886 the U.S. Army began enforcing the laws within the park. In 1916 the NPS was formed, complete with forest rangers trained in law enforcement.

The most notable features of Yellowstone are the thermal geysers, hot springs, and bubbling mud volcanoes. In all more than 10,000 thermal vents of one kind or another exist within the park, including over 300 geysers. Most of the geysers erupt only occasionally, but others spout regularly. The famous Old Faithful gushes hot water 130 feet into the air about every 70 minutes. Beautiful lakes, streams, and forests make Yellowstone an ideal spot for hiking, canoeing, camping, and fishing.

Yosemite National Park, California
P.O. Box 577
Yosemite National Park, CA 95389
(209) 372-0200
Established: 1890
Total acres: 761,170
Wilderness area acres: 677,600

It was primarily the efforts of John Muir that led to the establishment of Yosemite National Park, providing protection to Muir's beloved Yosemite

Valley, as well as surrounding valleys, meadows, streams, lakes, forests, and peaks within the Sierra Nevada Mountains. Yosemite Valley itself is the most stunning area of the park, with sheer granite monoliths rising straight up from the valley floor, including Half Dome and El Capitan, with a 3,500-foot sheer face. Upper Yosemite Falls plunges 1,430 feet into the valley.

Elsewhere the park contains a diverse set of ecosystems, from 2,000 feet above sea level to the 13,114-foot Mount Lyell. About 80 different species of mammals live within this range, from deer and black bears, to marmots. There are also more than 200 species of birds, 25 species of reptiles, and 9 species of amphibians.

Zion National Park, Utah
Springdale, UT 84767
(801) 772-3256
Established: 1919
Total acres: 147,035

Zion Canyon, named by Mormon missionaries after a mythical city of God, contains fantastic sheer red, gold, and orange rock cliffs dropping down to a valley carved by the Virgin River. Along the rim grow forests of Douglas and White Fir. Zion is considered by many to be a sister park to nearby Grand Canyon and Bryce Canyon national parks, but like the other two, it has a beauty and character all its own.

National Monuments

In many respects national monuments are similar to national parks. Many national parks were actually classified as monuments prior to becoming parks. While the parks protect natural sites almost exclusively, however, the national monuments protect natural, historic, and scientific areas. Some national monuments also contain wilderness areas.

Agate Fossil Beds National Monument, Nebraska
P.O. Box 427
Gering, NE 69341
(308) 668-2211
Established: 1965
Total acres: 3,055

The fossilized bones of prehistoric animals are found in abundance at this monument, including Menoceras, a two-horned rhino, and Moropus, an odd animal with characteristics of a bear, horse, and giraffe.

Alibates Flint Quarries National Monument, Texas
Box 1438
Fritch, TX 79036
(806) 857-3151
Established: 1965
Total acres: 1,371

Indians used these quarries for thousands of years to find highly prized flint for use in weapons and tools. The flint from these particular quarries was both hard and beautiful, with colorful rainbow hues and stripes. It was widely traded and has been located at archeological sites in New Mexico, Texas, Oklahoma, Kansas, Colorado, Arizona, Arkansas, and Louisiana. The monument protects over 250 quarry pits, which are shallow depressions in the ground 4 to 8 feet wide.

Aniakchak National Monument, Alaska
P.O. Box 7
King Salmon, AK 99613
(907) 246-3305
Established: 1978
Total acres: 139,500

Located 400 miles southwest of Anchorage, Aniakchak is a dormant volcano, last active in 1931. Surprise Lake is located inside the crater and has an average diameter of six miles. Water from the lake escapes through a gap in the crater wall to create the Aniakchak River.

Aztec Ruins National Monument, New Mexico
P.O. Box 640
Aztec, NM 87410
(505) 334-6174

These ruins were built by the Anasazi Indians in northern New Mexico and occupied between A.D. 1100 and 1300. The name came from early white settlers who mistakenly thought the site was built by Aztecs. The largest building here is a 500-room dwelling that measures 360 by 275 feet, and may have housed 450 people. The structure is made of mud bricks on a stone foundation and rises to four stories high.

Bandelier National Monument, New Mexico
Los Alamos, NM 87544
(505) 672-3861
Established: 1916
Total acres: 36,917
Wilderness area acres: 23,267

Ruins of Pueblo Indian cliff dwellings are preserved at this monument. Exhibits explain the ancient native lifestyle and self-guided tours lead through the ruins, many of which have not been excavated.

Black Canyon of the Gunnison National Monument, Colorado
P.O. Box 1648
Montrose, CO 81402
(303) 240-6522
Established: 1933
Total acres: 20,763
Wilderness area acres: 11,180

This canyon is carved by the wild Gunnison River deep into black and gray schist and gneiss. It is both deep (2,500 feet) and narrow. At its narrowest point the North and South rims are only 1,300 feet apart and the base is only 40 feet wide. The monument follows the most dramatic portion of the canyon for 12 miles.

Booker T. Washington National Monument, Virginia
Route 1, Box 195
Hardy, VA 24101
(703) 721-2094
Established: 1956
Total acres: 223

The birthplace of Booker T. Washington is re-created here in its original location. Washington went from a childhood in slavery to rise up to become one of the most influential African Americans of his time.

Buck Island Reef National Monument, Virgin Islands
Box 160
Christiansted
St. Croix, Virgin Islands 00820
(809) 773-1460
Total acres: 880

The 175-acre Buck Island is surrounded by a spectacular coral reef, filled with colorful angelfish, sea turtles, rays, and other aquatic life. The island is covered with tropical rainforest, uninhabited, and located 1.5 miles north of St. Croix.

Cabrillo National Monument, California
P.O. Box 6670
San Diego, CA 92106
(619) 293-5450
Established: 1913
Total acres: 144

Named as a tribute to Spanish explorer Juan Rodriguez Cabrillo, this monument includes a statue of Cabrillo, Point Loma lighthouse, a whale watching overlook, and a tidepool area.

Canyon de Chelly National Monument, Arizona
P.O. Box 588
Chinle, AZ 86503
(602) 674-5436
Established: 1931
Total acres: 83,840

Within this steep-walled canyon in Arizona lie the ruins of ancient Native American villages, built between A.D. 350 and 1300. Navajos still live in the Canyon today. The monument includes scenic drives and trails, and archeology walks.

Cape Krusenstern National Monument, Alaska
P.O. Box 287
Kotzebue, AK 99752
(907) 442-3890
Established: 1980
Total acres: 660,000

This undeveloped wilderness along the northwestern coast of Alaska is an ancient Eskimo hunting ground for marine mammals. Eskimo artifacts dating back 6,000 years can be found here, and native populations still return to hunt seals every summer. This monument is isolated, with the closest road 600 miles away.

Capulin Volcano National Monument, New Mexico
Capulin, NM 88414
(505) 278-2201
Established: 1916
Total acres: 793

Capulin is a 10,000-year-old extinct volcano. A trail leads down into the crater, through grass, trees, and black volcanic rocks. The peak itself is a symmetrical cone 1,000 feet above an 8,182-foot base.

Casa Grande National Monument, Arizona
P.O. Box 518
Coolidge, AZ 85228
(602) 723-7209
Established: 1889
Total acres: 473

The Casa Grande is a large, four-story structure built by the prehistoric Hohokum Indians in Arizona's Gila River Valley. Scientists are not sure what purpose the building had, but suspect that it may have been an astronomical observatory. Surrounding the structure are the ruins of ancient villages, abandoned sometime in the late 1300s.

Castillo de San Marcos National Monument, Florida
1 Castillo Drive
St. Augustine, FL 32084
(904) 829-6506
Established: 1924
Total acres: 20

Begun by the Spanish in 1672 and completed in 1695, Castillo de San Marcos is the oldest masonry fort in the United States. The fort was built to protect the Spanish settlement at St. Augustine from the British. St. Augustine was the first permanent European settlement in what is now the United States, founded in 1565.

Castle Clinton National Monument, New York
Manhattan Sites, National Park Service
26 Wall Street
New York, NY 10005
Established: 1946
Total acres: 1

On the tip of Manhattan this structure was built as a harbor defense fortification in 1808–1811. It was used as an immigration depot from 1855 to 1890. Today it is part of Battery Park.

Cedar Breaks National Monument, Utah
P.O. Box 749
Cedar City, UT 84720
(801) 586-9451
Established: 1933
Total acres: 6,155

Cedar Breaks is a natural amphitheater carved out of pink and purple limestone. Surrounding the rock formations are forests of fir, aspen, and bristlecone pine.

Chiricahua National Monument, Arizona
Don Cabeza Route, Box 6500
Willcox, AZ 85643
(602) 824-3560
Established: 1924
Total acres: 11,985
Wilderness area acres: 9,440

Ancient volcanic rock has eroded over time to create a wonderland of odd formations at the base of the Chiricahua Mountains. The mountain's northern slope is lush with sycamore, cypress, and oak trees. On the south side is typical southwestern desert vegetation.

Colorado National Monument, Colorado
Fruita, CO 81521
(303) 858-3617
Established: 1911
Total acres: 20,454

Colorado National Monument contains canyons, fantastic rock formations, Native American relics and petroglyphs, and dinosaur fossils. Trails lead past the more interesting formations and into the backcountry for overnight trips.

Congaree Swamp National Monument, South Carolina
P.O. Box 11938
Columbia, SC 29211
(803) 765-5571
Established: 1976
Total acres: 22,000
Wilderness area acres: 15,000

This swamp forest is only a 30-minute drive from downtown Charleston, South Carolina. It contains 85 species of trees, from cypress, tupelo, and elms in the soggier areas, to oak, maple, hickory, and willow in the drier parts.

Craters of the Moon National Monument, Idaho
P.O. Box 20
Arco, ID 83213
(208) 527-3257
Established: 1924
Total acres: 53,454
Wilderness area acres: 43,243

The striking lunar-like landscape at Craters of the Moon National Monument was formed as molten rock was forced to the surface during the past 15,000 years. Astronauts trained here in preparation for visits to the moon. Big Cinder Butte is the largest cinder cone in the world.

Devils Postpile National Monument, California
c/o Sequoia and Kings Canyon National Parks
Three Rivers, CA 93271
(209) 561-3314
Total acres: 798

The 60-foot-high hexagonal basalt columns at Devils Postpile were formed when lava cracked as it cooled 100,000 years ago. Located in the Sierra Nevada Mountains, the sheer vertical columns measure two to three feet in diameter.

Devils Tower National Monument, Wyoming
Devils Tower, WY 82714
(307) 467-5370
Established: 1906
Total acres: 1,347

Basalt columns, similar in appearance to Devils Postpile, formed Devils Tower in Wyoming 60 million years ago. This volcanic edifice rises 867 feet high. It was the first national monument established in the nation.

Dinosaur National Monument, Colorado–Utah
Box 210
Dinosaur, CO 81610
(303) 374-2216
Established: 1915
Total acres: 211,142

Fossilized dinosaur bones found here include those from the Stegosaurus, Brontosaurus, and Allosaurus. It is the largest Jurassic Period dinosaur quarry currently known. In 1950, plans were introduced to build a dam and flood the valley, but after a considerable battle the idea was defeated.

Effigy Mounds National Monument, Iowa
McGregor, IA 52157
(319) 873-2356
Established: 1949
Total acres: 1,481

Native American burial grounds are preserved here in northeastern Iowa. The 191 mounds include numerous ornaments, spearpoints, and other relics. Twenty-six of them are shaped like birds or bears.

El Malpais National Monument, New Mexico
P.O. Box 939
Grants, NM 87020
(505) 285-5406
Established: 1987
Total acres: 114,716

El malpais means "the badlands" in Spanish. This monument consists of a valley of lava flows from 3- to 700-million years old, including the longest and largest lava tube system in North America. These caves stretch more than 16 miles long and up to 100 feet in diameter. Ancient Native American roads and ruins, sandstone bluffs, and forested areas are also within the monument.

El Morro National Monument, New Mexico
Rahah, NM 87321
(505) 783-5132
Established: 1906
Total acres: 1,279

Poking up from the surrounding desert is a striking 200-foot sandstone monolith with historic inscriptions from American Indians, Spanish, and early American explorers. *El morro* means "the knoll" in Spanish.

Florissant Fossil Beds National Monument, Colorado
P.O. Box 185
Florissant, CO 80816
(303) 748-3253
Established: 1969
Total acres: 5,998

More than 80,000 fossils have been found here, as the remains of plants and insects became embedded in lake bottom silt 35 million years ago.

Fort Frederica National Monument, Georgia
Route 9, Box 286-C
Saint Simons Island, GA 31522
(912) 638-3639
Established: 1936
Total acres: 216

Built around 1736, this fort served the British in their competition with Spain for dominance in North America. In 1739 it was used as the base for a British attack against the Spanish Castle of St. Augustine, Florida. In 1742 the Spanish attacked nearby St. Simons. By 1748 most of the British troops were withdrawn and the fort fell into disrepair.

Fort McHenry National Monument and Historic Shrine, Maryland
Baltimore, MD 21230
(301) 962-4290
Established: 1925
Total acres: 43

During the War of 1812, Frances Scott Key was captured by the British and held on a ship in Baltimore Harbor. It was here that he witnessed the 25-hour bombardment of Fort McHenry by the British. In the morning, when he saw that the U.S. flag was still waving, he penned the words to what would become the national anthem. The Fort McHenry National Monument commemorates the battle of Baltimore.

Fort Matanzas National Monument, Florida
1 Castillo Drive
St. Augustine, FL 32084
(904) 471-0016
Established: 1924
Total acres: 228

After a battle in 1565, 500 French soldiers surrendered to the Spanish here, but all but 31 were executed. This fort was built by the Spanish at St. Augustine in 1740–1742 to guard against the British. *Matanzas* means "slaughterers" in Spanish.

Fort Pulaski National Monument, Georgia
Box 98
Tybee Island, GA 31328
(912) 786-5787
Established: 1924
Total acres: 5,623

Fort Pulaski was originally deemed to be impregnable, but Union cannons proved that to be incorrect early in the Civil War. The fort was severely damaged in April 1862, forcing the Confederate troops inside to surrender.

Fort Stanwix National Monument, New York
112 East Park Street
Rome, NY 13440
(315) 336-2090
Established: 1935
Total acres: 16

The log and earth Fort Stanwix was built by the British in 1758. It was the site of an important battle during the Revolutionary War, during which Benedict Arnold, still a patriot, tricked Iroquois Indians into deserting their British allies by sending a message that a huge number of colonial reinforcements was approaching.

Fort Sumter National Monument, South Carolina
1214 Middle Street
Sullivan's Island, SC 29482
(803) 883-3123
Established: 1948
Total acres: 197

Fort Sumter was the site of the first battle in the Civil War, which began on 12 April 1861. The monument also includes Fort Moultrie on the other side of the harbor.

Fort Union National Monument, New Mexico
Watrous, NM 87753
(506) 425-8025
Established: 1954
Total acres: 721

At its height, Fort Union was the largest fort west of the Mississippi. Three forts were actually built in the site; the first of wood in 1851, and the second and third of adobe during the 1860s. A total of 50 buildings were constructed there, but only weathered adobe ruins remain.

Fossil Butte National Monument, Wyoming
Box 527
Kemerer, WY 83101
(307) 877-3450
(307) 877-5500
Established: 1969
Total acres: 8,198

An extensive array of 50-million-year-old freshwater fish are fossilized here in an 18-inch-thick layer of Green River Limestone. Species include stingray, perch, herring, paddlefish, and various insects. Other layers have included horses, monkeys, and crocodiles. The monument is covered with grasslands and is home to an abundance of wildlife, from deer, antelope, and elk to bobcats and coyotes.

George Washington Birthplace National Monument, Virginia
Rural Route 1, Box 717
Washington's Birthplace, VA 22575
(804) 224-1732
Established: 1930
Total acres: 538

The plantation on which George Washington was born is re-created here in its original spot, including the manor and surrounding tobacco farm. Washington lived here until he was 3-1/2 years old.

George Washington Carver National Monument, Missouri
P.O. Box 38
Diamond, MO 64840
(417) 325-4151
Established: 1943
Total acres: 210

George Washington Carver was born at this site as a black slave and lived here until he was 12 years old. Later, as a scientist at the Tuskegee Institute in Alabama, Carver discovered over 300 uses for the peanut and 100 uses for the sweet potato.

Gila Cliff Dwellings National Monument, New Mexico
Route 11, Box 100
Silver City, NM 88061
(505) 536-9461
Established: 1907
Total acres: 533

Located within the Gila National Forest and the Gila Wilderness Area, this monument contains the ruins of masonry dwellings built into caves and cliffs by the Mogollon Indians in the late twelfth and early thirteenth centuries. The mountains surrounding the ruins are home to a variety of wildlife, from deer, elk, mountain lions, foxes, and bears to hawks and eagles.

Grand Portage National Monument, Minnesota
P.O. Box 666
Grand Marais, MN 55604
(218) 387-2788
Established: 1958
Total acres: 710

On the western slope of Lake Superior, this spot was the main trading post for fur trappers during the late 1700s. A collection of wood and log buildings still stands on the spot.

Great Sand Dunes National Monument, Colorado
11500 Highway 150
Mosca, CO 81146
(719) 378-2312
Established: 1932
Total acres: 38,662
Wilderness area acres: 33,450

These sand dunes at the base of the Sangre de Cristo Mountains in south central Colorado cover an area of 155 square miles. The wind can blow up to 120 miles per hour here, constantly shaping the dunes, and in the summer the sand can heat up to 140 degrees Fahrenheit.

Hagerman Fossil Beds National Monument, Idaho
2647 Kimberly Road East
Twin Falls, ID 83301
(208) 733-8398
Established: 1988
Total acres: 4,394

Pliocene era fossils, from 2.5 to 3.5 million years old are found here, including many small rodents, birds, fish, amphibians, and reptiles. The site may be best known for the discovery of an ancient zebra-like horse, known now as the Hagerman Horse.

Hohokam Pima National Monument, Arizona
c/o Director of Economic Development
Gila River Indian Community
Sacaton, AZ 85247
(602) 562-3311
Established: 1972
Total acres: 1,690

Prehistoric Hohokam Indian ruins are located on the current Pima Indian Reservation here in Arizona, but although it is a national monument, the site is not open to non-Indian visitors. Other parts of the reservation are open, however, including a Gila River Indian Arts and Crafts Center.

Homestead National Monument of America, Nebraska
Route 3, Box 47
Beatrice, NE 68310
(402) 223-3514
Established: 1936
Total acres: 195

The Homestead National Monument commemorates the Homestead Act of 1862 and marks the spot of one of the first free land grants. A brick school house built by settlers Daniel and Agnes Freeman still stands on the spot.

Hovenweep National Monument, Colorado–Utah
McElmo Route
Cortez, CO 81321
No phone
Established: 1923
Total acres: 785

Six sets of 700-year-old Anasazi Indian ruins are located here, with four in Colorado and two in Utah. Square and circular towers and 20-foot-high walls remain from the original pueblos.

Jewel Cave National Monument, South Dakota
Rural Route 1, Box 60AA 351
Custer, SD 57747
(605) 673-2288
Established: 1908
Total acres: 1,274

With 79 miles of underground caverns, Jewel Cave is the second longest cave in the United States; second only to Kentucky's Mammoth Cave. The site is located in South Dakota's Black Hills, and is the only national monument in South Dakota.

John Day Fossil Beds National Monument, Oregon
420 West Main Street
John Day, OR 97845
(503) 575-0721
Established: 1974
Total acres: 14,030

Plant and animal remains from four separate epochs representing over 45 million years are fossilized here, spread over three sites. The John Day River passes through the monument and scenic trails lead through fir and pine forests.

Lava Beds National Monument, California
P.O. Box 867
Tulelake, CA 96134
(916) 667-2282
Established: 1925
Total acres: 46,560
Wilderness area acres: 28,460

Cinder cones, lava flows, and caves make up this monument in northern California. More than 200 lava tube caves are open for exploration by visitors. In 1872, 54 Modoc Indians used the harsh terrain to hold off 1,000 U.S. troops in a six-month long siege that ended in their deaths.

Little Bighorn Battlefield National Monument, Montana
(formerly Custer Battlefield National Monument)
P.O. Box 39
Crow Agency, MT 59022
(406) 638-2622
Established: 1946
Total acres: 765

It was here, in 1876, that Sioux and Cheyenne Indians joined together to annihilate the troops of General George Armstrong Custer. None of the U.S. soldiers survived the battle. The monument is located on two adjacent tracts of land on the plains of Montana.

Montezuma Castle National Monument, Arizona
P.O. Box 219
Camp Verde, AZ 86322
(602) 567-3322

Established: 1906
Total acres: 858

Montezuma Castle is a 20-room dwelling built into a cliffside by Sinagua Indians about 800 years ago. It was named by white settlers who mistakenly believed it was built by Aztecs. Visitors are not allowed to enter the structure, or even climb the cliffs to its entrance, but can view it from a facing canyon rim.

Muir Woods National Monument, California
Mill Valley, CA 94941
(415) 388-2595
Established: 1908
Total acres: 553

Conservationist William Kent purchased this wooded valley and donated it to the U.S. government in order to preserve redwood forests, as well as the wildlife that lives within them. The monument was named in honor of John Muir.

Natural Bridges National Monument, Utah
Box 1
Lake Powell, UT 84533
(801) 259-7164
Established: 1908
Total acres: 7,636

Three gigantic, naturally carved stone bridges are the centerpiece of this impressive monument. Owachomo bridge is 106 feet high by 180 feet long, Kachina bridge is 210 feet by 204 feet, and Sipapu bridge is 220 feet by 268 feet. Anasazi Indian ruins are also located in spots throughout the monument.

Navajo National Monument, Arizona
Tonalea, AZ 86044
(602) 672-2366
Established: 1909
Total acres: 360

Some of the largest and best preserved Anasazi Indian cliff dwellings are located here. Rangers give guided tours to limited numbers of visitors each day. The Anasazi and Navajo tribes are not related, but the monument is surrounded by the Navajo Indian Reservation.

Ocmulgee National Monument, Georgia
1207 Emery Highway
Macon, GA 31201
(912) 752-8257

Established: 1934
Total acres: 683

The Mississippian Indians lived at this site from 900 to 1100. Several mounds and part of a public lodge built 1,000 years ago still stand on the spot.

Oregon Caves National Monument, Oregon
19000 Caves Highway
Cave Junction, OR 97523
(503) 592-2100
Established: 1909
Total acres: 488

Visitors to Oregon Caves National Monument can walk through the caves past interesting marble formations, stalactites, and stalagmites. Outside, another trail runs through forests of Douglas fir, one of which is 1,500 years old.

Organ Pipe Cactus National Monument, Arizona
Route 1, Box 100
Ajo, AZ 85321
(602) 387-6849
Established: 1937
Total acres: 1,373
Wilderness area acres: 312,600

The Organ Pipe Cactus National Monument preserves a portion of Sonoran Desert ecosystem along the Arizona–Mexico border. The rare organ pipe cactus grows here in profusion, along with other cactus varieties, such as the cholla and saguaro, which exists only in the Sonoran Desert.

Petroglyph National Monument, New Mexico
c/o Southwest Regional Office
National Park Service
P.O. Box 728
Santa Fe, NM 87504
(505) 988-6375
Established: 1990
Total acres: 5,188

More than 15,000 prehistoric American Indian and historic Spanish petroglyphs are carved or painted on the rocks of a 17-mile stretch of Albuquerque's West Mesa Escarpment. More than 60 archeological sites also exist on the monument.

Pinnacles National Monument, California
Paicines, CA 95043
(408) 389-4485
Established: 1908
Total acres: 16,265
Wilderness area acres: 12,952

Volcanoes formed the rock outcroppings, spires, and pinnacles at this monument, some of which rise 600 feet from the ground. The area is a favorite for rock climbers, and includes 26 miles of hiking trails.

Pipe Spring National Monument, Arizona
Moccasin, AZ 86022
(602) 643-7105
Established: 1923
Total acres: 40

Pipe Spring is an oasis in the Arizona desert where Mormon settlers started a cattle ranch in 1870. The monument was established to honor the early settlers and itinerant cowboys of that era. A small fort, corrals, and other original buildings still stand on the site.

Pipestone National Monument, Minnesota
P.O. Box 727
Pipestone, MN 56164
(507) 825-5464
Established: 1937
Total acres: 283

Native American tribes from across North America traveled to this spot, beginning about 400 years ago, in order to dig stone from quarries considered sacred. The stone was thought to be made from the same material as the Indians themselves, and was used to make ceremonial pipes, often referred to as "peace pipes."

Poverty Point National Monument, Louisiana
HC 60, Box 208A
Epps, LA 71237
(318) 926-5492
Established: 1988
Total acres: 911

Archaeologists estimate that between 4,000 and 5,000 American Indians lived here in a prehistoric city between 1800 and 500 B.C. Six concentric horseshoe-shaped mounds and a 70-foot-high by up to 800-foot-wide bird-shaped mound remain.

Rainbow Bridge National Monument, Utah
P.O. Box 1507
Page, AZ 86040
(602) 645-2471
Established: 1910
Total acres: 160

Rainbow Bridge is the largest natural bridge in the world at 290 feet high and 278 feet long. The bridge was carved by a river out of salmon-pink colored sandstone.

Russell Cave National Monument, Alabama
Route 1, Box 175
Bridgeport, AL 35740
(205) 495-2672
Established: 1961
Total acres: 310

American Indians spent winters living in Russell Cave in Alabama for 9,000 years. Today two tons of archeological artifacts have been recovered from the cave, which measures 107 feet wide, by 258 feet deep, by 29 feet high.

Saguaro National Monument, Arizona
3693 S. Old Spanish Trail
Tuscon, AZ 85730
East (602) 296-8576
West (602) 883-6366
Established: 1933
Total acres: 83,574
Wilderness area acres: 71,400

Saguaro National Monument includes two separate areas, one on the east and the other on the west of Tuscon, Arizona. Each area contains large stands of saguaro cactus, which are found only in the Sonoran Desert. These cactus are often associated with the desert in popular art and film. They are round and tall, growing to a height of 50 feet.

Salinas Pueblo Missions National Monument, New Mexico
P.O. Box 496
Mountainair, NM 80736
(505) 847-2585
Established: 1909
Total acres: 1,077

The Spanish built several missions on this site next to Mogollon Indian pueblos, beginning in 1629. In the 1670s the Indians left the area, and today the Spanish churches are in ruins.

Scotts Bluff National Monument, Nebraska
P.O. Box 427
Gering, NE 69341
(308) 436-4340
Established: 1919
Total acres: 2,998

Scotts Bluff was a famous landmark for pioneers on the Oregon Trail during the 1800s. The bluff rises 800 feet above the otherwise flat prairie. The monument preserves both the bluff and a segment of the original Oregon Trail.

Statue of Liberty National Monument, New York–New Jersey
Liberty Island
New York, NY 10004
(212) 363-3200
Established: 1924
Total acres: 58

The Statue of Liberty in New York Harbor stands 111 feet high above the base and weighs 225 tons. It is made of copper and was a gift from France in 1884. It was designed by the French sculptor Frederic Auguste Bartholdi and engineer Alexandre-Gustave Eiffel.

Sunset Crater Volcano National Monument, Arizona
Route 3, Box 149
Flagstaff, AZ 86004
(602) 527-7042
Established: 1930
Total acres: 3,040

This monument includes the 3,000-acre Sunset Crater, north of Flagstaff, and a 225-foot-long lava cave through one of the lava flows.

Timpanogos Cave National Monument, Utah
Route 3, Box 200
American Fork, UT 84003
(801) 756-5239
Established: 1922
Total acres: 250

Three large caves are connected by man-made tunnels underneath the 11,750-foot Mount Timpanogos in the Wasatch Mountains. These caves are well known for their helictite mineral formations.

Tonto National Monument, Arizona
P.O. Box 707
Roosevelt, AZ 85545
(602) 467-2241
Established: 1907
Total acres: 1,120

These cliff dwellings were built by the Salado Indians around 600 years ago. Some of the ruins here are preserved intact and can be explored by visitors.

Tuzigoot National Monument, Arizona
P.O. Box 219
Camp Verde, AZ 86322
(602) 634-5564
Established: 1939
Total acres: 801

The ruins of a 110-room dwelling built by the Sinagua Indians sit here on a ridge above the Verde Valley. The building was originally two to three stories high and was constructed between 1125 and 1400.

Walnut Canyon National Monument, Arizona
Walnut Canyon Road
Flagstaff, AZ 86004
(602) 526-3367
Established: 1915
Total acres: 2,249

The ruins of 24 cliff dwellings built by the Sinagua Indians during the twelfth century can be found here. This lush canyon supports a wide variety of plants and animals, but no one knows why the inhabitants abandoned their villages after living in them for less than 150 years.

White Sands National Monument, New Mexico
P.O. Box 458
Alamagordo, NM 88310
(505) 479-6134
Established: 1933
Total acres: 143,733

The gleaming white sand in this 230-square-mile stretch of dunes is actually gypsum. This is the largest gypsum field in the world, and some of the dunes rise 45 feet high. The monument is entirely surrounded by the White Sands Missile Range, which was the site of the first atomic bomb explosion, conducted on 16 July 1945.

Wupatki National Monument, Arizona
HC 33, Box 444A
Flagstaff, AZ 86001
(602) 774-7000
Established: 1924
Total acres: 35,253

More than 200 Sinagua Indian ruins are located within this monument north of Flagstaff. Most notable are a 100-room dwelling, an amphitheater, and a ball court.

Yucca House National Monument, Colorado
c/o Mesa Verde National Park
Mesa Verde, CO 81330
(303) 529-4465
Established: 1919
Total acres: 10

All that is left of an ancient Native American village that once stood on this spot is an assortment of mounds, the largest of which is 15 to 20 feet high. The site has not yet been excavated, but it is thought that the mounds will yield interesting clues to the lives of the original inhabitants.

National Lakeshores

National lakeshores protect two shoreline areas on Lake Michigan and two on Lake Superior.

Apostle Islands National Lakeshore, Wisconsin
Established: 1970   Total acres: 69,372

Indiana Dunes National Lakeshore, Indiana
Established: 1966   Total acres: 12,857

Pictured Rocks National Lakeshore, Michigan
Established: 1966   Total acres: 72,899

Sleeping Bear Dunes National Lakeshore, Michigan
Established: 1970   Total acres: 71,132

National Preserves

The national preserves are lands set aside to preserve natural resources. The majority of these areas are in the Alaskan wilderness. Denali, Gates of the Arctic, Glacier Bay, Katmai, Lake Clark,
and Wrangell–St. Elias are connected to national parks of the
same name. Aniakchak borders Aniakchak National Monument.

Aniakchak National Preserve, Alaska
Established: 1980   Total acres: 475,500

Bering Land Bridge National Preserve, Alaska
Established: 1980   Total acres: 2,784,960

Big Cypress National Preserve, Florida
Established: 1974   Total acres: 716,000

Big Thicket National Preserve, Texas
Established: 1974   Total acres: 85,733

Denali National Preserve, Alaska
Established: 1980   Total acres: 1,311,365

Gates of the Arctic National Preserve, Alaska
Established: 1980   Total acres: 948,629

Glacier Bay National Preserve, Alaska
Established: 1980   Total acres: 57,844

Katmai National Preserve, Alaska
Established: 1980   Total acres: 374,000

Lake Clark National Preserve, Alaska
Established: 1980   Total acres: 1,407,293

Little River Canyon National Preserve, Alabama
Established: 1992   Total acres: 13,669

Noatak National Preserve, Alaska
 Established: 1980   Total acres: 6,574,481

Timucuan Ecological and Historic Preserve, Florida
 Established: 1988   Total acres: 4,856,721

Wrangell–St. Elias National Preserve, Alaska
 Established: 1980   Total acres: 4,856,721

Yukon Charley Rivers National Preserve, Alaska
 Established: 1980   Total acres: 2,523,509

National Rivers

National rivers are rivers of historic or scenic importance, though they do not have to be wild. In other words, dams can exist on the national rivers.

Big South Fork National River and Recreation Area, Tennessee–Kentucky
 Established: 1974   Total acres: 122,960

Buffalo National River, Arkansas
 Established: 1972   Total acres: 94,219
            Wilderness area acres: 10,529

Mississippi National River and Recreation Area, Minnesota
 Established: 1988   Miles: 69

New River Gorge National River, West Virginia
 Established: 1978   Total acres: 62,663

Niobrara National Scenic River, Nebraska–South Dakota
 Authorized: 1991

Ozark National Scenic Riverways, Missouri
 Established: 1972   Total acres: 80,788

National Scenic Trails

The National Trails System includes 16 national scenic and historic trails administered by both the Agriculture and Interior Departments. Only the following three are units of the National Park System:

Appalachian National Scenic Trail, Maine to Georgia
Established: 1968   Miles: 2,100

Natchez Trace National Scenic Trail, Mississippi to Tennessee
Established: 1983   Miles: 694

Potomac Heritage National Scenic Trail, Virginia to Pennsylvania
Established: 1983   Miles: 704

National Seashores

National seashores are scenic coastal areas that are protected in a pristine, undeveloped state.

Assateague Island National Seashore, Maryland–Virginia
Established: 1965   Total acres: 39,631

Canaveral National Seashore, Florida
Established: 1975   Total acres: 57,662

Cape Cod National Seashore, Massachusetts
Established: 1961   Total acres: 43,557

Cape Hatteras National Seashore, North Carolina
Established: 1937   Total acres: 30,319

Cape Lookout National Seashore, North Carolina
Established: 1966   Total acres: 28,415

Cumberland Island National Seashore, Georgia
Established: 1972   Total acres: 36,415
                                     Wilderness area acres: 8,840

Fire Island National Seashore, New York
Established: 1964   Total acres: 19,579
                                     Wilderness area acres: 1,363

Gulf Islands National Seashore, Florida–Mississippi
Established: 1971   Total acres: 65,817
                                     Wilderness area acres: 1,800

Padre Island National Seashore, Texas
Established: 1962   Total acres: 130,697

Point Reyes National Seashore, California
Established: 1962   Total acres: 71,046
                                     Wilderness area acres: 25,370

National Wild and Scenic Rivers

Like national rivers, national wild and scenic rivers have historic
and scenic value, but they must also be wild, with no dams or other
control of the river's flow.

Alagnak Wild River, Alaska
Established: 1980   Miles: 69

Bluestone National Scenic River, West Virginia
Established: 1988   Miles: 11

Delaware National Scenic River, Pennsylvania–New Jersey–New York
Established: 1978   Miles: 41

Egg Harbor National Scenic and Recreational River, New Jersey
Established: 1994   Miles: 129

Missouri National Recreation River, Nebraska–South Dakota
Established: 1972   Miles: 27

Obed Wild and Scenic River, Tennessee
Established: 1978   Miles: 45

Rio Grande Wild and Scenic River, Texas
Established: 1978   Miles: 191

Saint Croix National Scenic Riverway, Minnesota–Wisconsin
Established: 1968   Miles: 227

**Upper Delaware Scenic and Recreational River,
New York–Pennsylvania**
Established: 1978   Miles: 73.4

National Wildlife Refuges

The following table lists all of the national wildlife refuges, which are administered by the U.S. Fish and Wildlife Service (USFWS) to protect wildlife habitats. About three-quarters of these areas were designed primarily to protect waterfowl. While special protection is not granted to the refuges, some areas are protected because of a wilderness area designation. The units below that do not have wilderness area acres listed are not protected from commercial development and are permitted to have campgrounds, cabins, lodges, and other facilities.

Unit Name	Wilderness Area Acres	Total Acres
ALABAMA		
Blowing Wind Cave	0	264
Bon Secour	0	4,629
Choctaw	0	4,218
Eufaula	0	7,953
Fern Cave	0	199
Grand Bay	0	840

Unit Name	Wilderness Area Acres	Total Acres
ALABAMA		
Watercress Darter	0	7
Wheeler	0	34,170
FH Interest*	0	605
ALASKA		
Alaska Maritime	2,570,754	3,435,639
Alaska Peninsula	0	3,500,000
Arctic	8,000,000	19,285,922
Becharof	400,000	1,200,018
Innoko	1,240,000	3,850,000
Izembek	300,000	303,094
Kanuti	0	1,430,002
Kenai	1,355,566	1,904,792
Kodiak	0	1,656,363
Koyukuk	400,000	3,550,000
Nowitna	0	1,560,000
Selawik	240,000	2,150,001
Tetlin	0	700,054
Togiak	2,270,000	4,097,430
Yukon Delta	1,900,000	19,131,645
Yukon Flats	0	8,630,000
ARIZONA		
Buenos Aires	0	113,940
Cabeza Prieta	803,418	860,000
Cibola	0	8,606
Havazu	14,606	36,335
Imperial	9,220	17,810
Kofa	516,200	666,480
San Bernardino	0	3,608
ARKANSAS		
Big Lake	2,144	11,036
Cache River	0	21,873
Felsenthal	0	64,902
Holla Bend	0	6,077
Logan Cave	0	124
Overflow	0	11,404
Wapanocca	0	5,484
White River	0	113,230
FH Interest	0	3,459
CALIFORNIA		
Antioch Dunes	0	55
Bitter Creek	0	13,977

Unit Name	Wilderness Area Acres	Total Acres
CALIFORNIA		
Blue Ridge	0	897
Butte Sink	0	9,164
Castle Rock	0	14
Cibola	0	3,647
Clear Lake	0	33,440
Coachella Valley	0	3,074
Colusa	0	4,040
Delevan	0	5,634
Ellicott Slough	0	127
Farallon	141	211
Grasslands	0	42,518
Havasu	3,195	7,235
Hopper Mountain	0	2,471
Humboldt Bay	0	2,110
Imperial	5,836	7,958
Kern	0	10,618
Kesterson	0	11,500
Lower Klamath	0	40,294
Marin Islands	0	9
Merced	0	2,563
Modoc	0	6,386
North Central Valley	0	7,126
Pixley	0	6,192
Sacramento	0	10,783
Sacramento River	0	5,640
Salinas River	0	367
Salton Sea	0	37,579
San Francisco Bay	0	18,560
San Joaquin River	0	777
San Luis	0	8,133
San Pablo Bay	0	13,190
Seal Beach	0	911
Sutter	0	2,590
Sweetwater Marsh	0	316
Tijuana Slough	0	1,023
Tule Lake	0	39,199
Willow Creek–Lurline	0	4,596
FH Interest	0	80
COLORADO		
Alamosa	0	11,169
Arapaho	0	18,254
Browns Park	0	13,455

Unit Name	Wilderness Area Acres	Total Acres
COLORADO		
Monte Vista	0	14,189
Two Ponds Wetland Preserve	0	25
FH Interest	0	159
CONNECTICUT		
Stewart B. McKinney	0	347
DELAWARE		
Bombay Hook	0	15,122
Prime Hook	0	9,701
FH Interest	0	3
FLORIDA		
Archie Carr	0	13
Arthur R. Marshall Loxaha	0	145,665
Caloosahatchee	0	40
Cedar Keys	379	721
Chassahowitzka	23,579	30,436
Crocodile Lake	0	6,559
Crystal River	0	46
Egomont Key	0	328
Florida Panther	0	23,379
Great White Heron	1,900	7,408
Hobe Sound	0	980
Island Bay	20	20
J. N. Ding Darling	2,619	5,348
Key West	2,019	2,019
Lake Woodruff	1,066	19,545
Lower Suwannee	0	50,139
Matlacha Pass	0	512
Merritt Island	0	138,263
National Key Deer	2,278	8,110
Okefenokee	0	3,678
Passage Key	36	64
Pelican Island	6	4,426
Pine Island	0	548
Pinellas	0	392
St. Johns	0	6,255
St. Marks	17,350	65,400
St. Vincent	0	12,490
FH Interest	0	1,456
GEORGIA		
Banks Lake	0	4,049
Blackbeard Island	3,000	5,618

Unit Name	Wilderness Area Acres	Total Acres
GEORGIA		
Bond Swamp	0	4,596
Eufaula	0	3,231
Harris Neck	0	2,762
Okefenokee	353,981	391,402
Piedmont	0	34,903
Savannah	0	11,324
Tybee	0	100
Wassaw	0	10,070
Wolf Island	5,126	5,126
FH Interest	0	3,908
HAWAII		
Hakalau Forest	0	15,481
Hanalei	0	917
Hawaiian Islands	0	254,418
Huleia	0	241
James C. Campbell	0	166
Kakahaia	0	45
Kilauea Point	0	188
Pearl Harbor	0	61
IDAHO		
Bear Lake	0	18,068
Camas	0	10,578
Deer Flat	0	11,265
Grays Lake	0	16,579
Kootenai	0	2,774
Minidoka	0	20,723
ILLINOIS		
Chautauqua	0	6,446
Crab Orchard	4,050	43,662
Cypress Creek	0	5,443
Mark Twain	0	16,579
Meredosia	0	2,188
Mississippi River Caue	0	20,120
Upper Mississippi River	0	3,300
FH Interest	0	739
INDIANA		
Muscatatuck	0	7,802
FH Interest	0	253
IOWA		
Desoto	0	3,503
Driftless Area	0	507

Unit Name	Wilderness Area Acres	Total Acres
IOWA		
Mark Twain	0	10,471
Mississippi River Caue	0	30,315
Union Slough	0	2,916
Upper Mississippi River	0	20,669
Walnut Creek	0	4,116
FH Interest	0	92
KANSAS		
Flint Hills	0	18,463
Kirwin	0	10,778
Marais Des Cygnes	0	5,101
Quivira	0	21,820
FH Interest	0	117
KENTUCKY		
Reelfoot	0	2,040
LOUISIANA		
Atchafalaya	0	15,255
Bayou Cocodrie	0	4,932
Bayou Sauvage	0	18,000
Bogue Chitto	0	28,671
Breton	5,000	9,047
Cameron Prairie	0	9,621
Catahoula	0	5,318
D'Arbonne	0	17,419
Delta	0	48,799
Grand Cote	0	6,077
Lacassine	3,346	32,625
Lake Ophelia	0	14,477
Sabine	0	139,437
Shell Keys	0	8
Tensas River	0	59,023
Upper Ouachita	0	20,890
FH Interest	0	12,882
MAINE		
Cross Island	0	1,703
Franklin Island	0	12
Moosehorn	7,392	23,914
Petit Manan	0	3,334
Pond Island	0	10
Rachel Carson	0	4,216
Seal Island	0	65

Unit Name	Wilderness Area Acres	Total Acres
MAINE		
Sunkhaze Meadows	0	9,337
FH Interest	0	622
MARYLAND		
Blackwater	0	17,860
Chinocoteague	0	418
Eastern Neck	0	2,286
Martin	0	4,423
Patuxent	0	12,276
Susquehanna	0	4
FH Interest	0	68
MASSACHUSETTS		
Great Meadows	0	3,370
Massasoit	0	184
Monomoy	2,420	2,702
Nantucket	0	40
Nomans Land Island	0	620
Oxbow	0	711
Parker River	0	4,652
Thacker Island	0	22
MICHIGAN		
Harbor Island	0	695
Huron	148	148
Kirtlands Warbler	0	6,310
Michigan Islands	12	363
Seney	25,150	95,455
Shiawassee	0	8,984
Wyandotte	0	304
FH Interest	0	2,520
MINNESOTA		
Agassiz	4,000	61,501
Big Stone	0	11,275
Hamden Slough	0	2,197
Mid Continent WMP	0	5,077
Mille Lacs	0	1
Minnesota Valley	0	7,949
Mississippi River Caue	0	15,421
Rice Lake	0	16,371
Rydell NWR	0	2,070
Sherburnie	0	29,606
Tamarac	2,180	35,167

Unit Name	Wilderness Area Acres	Total Acres
MINNESOTA		
Upper Mississippi River	0	18,083
FH Interest	0	1,521
MISSISSIPPI		
Bogue Chitto	0	6,808
Dahomey	0	4,850
Grand Bay	0	2,649
Hillside	0	15,406
Mathews Brake	0	2,418
Mississippi Sandhill Crane	0	19,303
Morgan Brake	0	4,865
Noxubee	0	46,673
Panther Swamp	0	28,600
St. Catherine Creek	0	13,035
Tallahatchie	0	3,898
Yazoo	0	12,940
FH Interest	0	19,735
MISSOURI		
Clarence Cannon	0	3,750
Mark Twain	0	1,352
Mingo	7,730	21,746
Ozark Cavefish	0	40
Pilot Knob	0	90
Squaw Creek	0	7,246
Swan Lake	0	11,348
FH Interest	0	567
MONTANA		
Benton Lake	0	12,453
Black Coulee	0	1,309
Bowdoin	0	15,552
Charles M. Russell	0	903,332
Creedman Coulee	0	2,728
Hailstone	0	920
Halfbreed Lake	0	4,318
Hewitt Lake	0	1,361
Lake Mason	0	16,660
Lake Thibadeau	0	3,868
Lamesteer	0	800
Lee Metcalf	0	2,793
Medicine Lake	11,366	31,484
National Bison Range	0	18,497
Nine-Pipe	0	2,022

Unit Name	Wilderness Area Acres	Total Acres
MONTANA		
Pable	0	2,542
Red Rock Lakes	32,350	44,158
Swan River	0	1,569
UI Bend	20,819	56,050
War Horse	0	3,192
FH Interest	0	511
NEBRASKA		
Crescent Lake	0	45,850
DeSoto	0	4,324
Fort Niobrara	4,635	19,133
Karl E. Mundt	0	19
North Platte	0	5,047
Valentine	0	71,517
FH Interest	0	790
NEVADA		
Anaho Island	0	248
Ash Meadows	0	13,231
Desert	0	1,588,819
Fallon	0	17,902
Moapa Valley	0	32
Pahranagat	0	5,383
Ruby Lake	0	37,631
Sheldon	0	570,461
Stillwater	0	78,086
NEW HAMPSHIRE		
Great Bay	0	1,054
John Hay	0	143
Wapack	0	1,672
NEW JERSEY		
Cape May	0	5,688
Edwin B. Forsythe	6,681	39,461
Great Swamp	3,660	7,238
Supawna Meadows	0	2,857
Wallkill River NWR	0	539
NEW MEXICO		
Bitter Lake	9,621	24,526
Bosque del Apache	30,287	57,191
Grulla	0	3,231
Las Vegas	0	8,672
Maxwell	0	3,699

Unit Name	Wilderness Area Acres	Total Acres
NEW MEXICO		
San Andres	0	57,215
Sevilleta	0	229,674
NEW YORK		
Amagansett	0	36
Conscience Point	0	60
Elizabeth A. Morton	0	187
Iroquios	0	10,821
Montezuma	0	6,446
Oyster Bay	0	3,204
Seatuck	0	209
Target Rock	0	80
Wertheim	0	2,424
FH Interest	0	1,856
NORTH CAROLINA		
Alligator River	0	141,596
Cedar Island	0	14,482
Currituck	0	2,046
Great Dismal Swamp	0	49,624
Mackay Island	0	6,997
Mattamuskeet	0	50,180
Pea Island	0	5,834
Pee Dee	0	8,438
Pocosin Lakes	0	107,719
Roanoke River	0	6,058
Swanquarter	8,785	16,411
FH Interest	0	4,323
NORTH DAKOTA		
Appert Lake	0	907
Ardoch	0	2,696
Arrowwood	0	15,934
Audubon	0	14,739
Bone Hill	0	640
Brumba	0	1,977
Buffalo Lake	0	2,096
Camp Lake	0	585
Canfield Lake	0	313
Chase Lake	4,155	4,385
Cottonwood	0	1,013
Dakota Lake	0	2,800
Des Lacs	0	19,547
Florence Lake	0	1,888

Unit Name	Wilderness Area Acres	Total Acres
NORTH DAKOTA		
Half-Way Lake	0	160
Hiddenwood	0	568
Hobart Lake	0	2,077
Hutchinson Lake	0	479
J. Clark Salyer	0	59,383
Johnson Lake	0	2,008
Kellys Slough	0	1,270
Lake Alice	0	11,355
Lake Elsie	0	635
Lake George	0	3,119
Lake Ilo	0	4,035
Lake Nettie	0	3,055
Lake Otis	0	320
Lake Zahl	0	3,823
Lambs Lake	0	1,207
Little Goose	0	288
Long Lake	0	22,499
Lords Lake	0	1,915
Lost Lake	0	960
Lostwood	5,577	26,904
McLean	0	760
Mapple River	0	712
Pleasant Lake	0	898
Pretty Rock	0	800
Rabb Lake	0	261
Rock Lake	0	5,506
Rose Lake	0	836
School Section Lake	0	680
Shell Lake	0	1,835
Sheyenne Lake	0	797
Sibley Lake	0	1,077
Silver Lake	0	3,348
Slade	0	3,000
Snyder Lake	0	1,550
Springwater	0	640
Stewart Lake	0	2,230
Stoney Slough	0	880
Storm Lake	0	686
Stump Lake	0	27
Sullys Hill	0	1,675
Sunburst Lake	0	327
Tewaukon	0	8,363
Tomahawk	0	440

Unit Name	Wilderness Area Acres	Total Acres
NORTH DAKOTA		
Upper Souris	0	32,302
White Lake	0	1,040
Wild Rice Lake	0	779
Willow Lake	0	2,620
Wintering River	0	239
Wood Lake	0	280
FH Interest	0	44
OHIO		
Cedar Point	0	2,445
Ottawa	0	5,793
West Sister Island	77	80
OKLAHOMA		
Little River	0	12,029
Oklahoma Bat Caves	0	593
Optima	0	4,333
Salt Plains	0	32,057
Sequoyah	0	20,800
Tishomingo	0	16,464
Washita	0	8,075
Wichita Mountains	8,570	59,020
OREGON		
Ankeney	0	2,796
Bandon Marsh	0	307
Basket Slough	0	2,492
Bear Valley	0	4,178
Cape Meares	0	139
Cold Springs	0	3,117
Deer Flat	0	162
Hart Mountain	0	251,295
Julia Butler Hansen	0	1,978
Klamath Forest	0	37,686
Lewis and Clark	0	38,172
Lower Klamath	0	6,618
Malheur	0	185,412
McKay Creek	0	1,837
Nestucca Bay	0	369
Oregon Islands	480	611
Sheldon	0	627
Siletz	0	40
Three Arch Rocks	15	15
Umatilla	0	8,880

Unit Name	Wilderness Area Acres	Total Acres
OREGON		
Upper Klamath	0	14,966
William L. Finley	0	5,332
FH Interest	0	358
PENNSYLVANIA		
Erie	0	8,750
John Heinz	0	923
Ohio River Islands	0	55
RHODE ISLAND		
Block Island	0	46
Ninigret	0	407
Pettaquamscutt Cove	0	154
Sachuest Point	0	242
Trustom Pond	0	642
SOUTH CAROLINA		
Ace Basin	0	2,861
Cape Romain	29,000	65,225
Carolina Sandhills	0	45,348
Pinckney Island	0	4,053
Santee	0	43,636
Savannah	0	14,285
FH Interest	0	501
SOUTH DAKOTA		
Bear Butte	0	375
Karl E. Mundt	0	1,603
Lacreek	0	16,855
Lake Andes	0	939
Pocasse	0	2,548
Sand Lake	0	21,818
Waubay	0	4,740
FH Interest	0	35
TENNESSEE		
Chickasaw	0	21,939
Cross Creeks	0	8,861
Hatchie	0	13,049
Lake Isom	0	1,846
Lower Hatchee	0	4,342
Reelfoot	0	8,408
Tennessee	0	51,358
FH Interest	0	541

Unit Name	Wilderness Area Acres	Total Acres
TEXAS		
Annahuac	0	28,560
Aransas	0	112,422
Attwater Prairie Chicken	0	7,984
Balcones Canyonlands	0	3,536
Big Boggy	0	4,526
Brazoria	0	42,338
Buffalo Lake	0	7,664
Grulla	0	5
Hagerman	0	12,142
Laguna Atascosa	0	45,187
Little Sandy	0	3,802
Lower Rio Grand Valley	0	60,709
McFaddin	0	42,956
Moody	0	3,517
Muleshoe	0	5,809
San Bernard	0	24,454
Santa Ana	0	2,088
Texas Point	0	8,952
UTAH		
Bear River	0	65,163
Fish Springs	0	17,992
Ouray	0	12,138
VERMONT		
Missisquoi	0	5,839
FH Interest	0	71
VIRGINIA		
Back Bay	0	5,568
Chincoteague	0	13,453
Eastern Shore of Virginia	0	651
Featherstone	0	326
Fisherman Island	0	1,025
Great Dismal Swamp	0	82,150
James River	0	4,147
Mackay Island	0	874
Marumsco	0	63
Mason Neck	0	2,276
Nansemond	0	208
Plum Tree Island	0	3,276
Presquile	0	1,329
Wallops Island	0	3,373
FH Interest	0	134

Unit Name	Wilderness Area Acres	Total Acres
WASHINGTON		
Columbia	0	29,597
Conboy Lake	0	5,814
Copalis	61	61
Dungeness	0	762
Flattery Rocks	125	125
Franz Lake	0	531
Grays Harbor	0	73
Julia Butler Hansen	0	2,777
Little Pend Orielle	0	39,999
McNary	0	3,631
Nisqually	0	2,847
Pierce	0	329
Protection Island	0	319
Quillayute Needles	300	300
Ridgefield	0	5,150
Saddle Mountain	0	30,818
San Juan Islands	353	449
Steigerwald Lake	0	627
Toppenish	0	1,979
Turnbull	0	17,824
Umatilla	0	14,676
Willapa	0	14,394
FH Interest	0	245
WEST VIRGINIA		
Ohio River Islands	0	311
WISCONSIN		
Fox River	0	838
Gravel Island	27	27
Green Bay	2	2
Horicon	0	21,176
Mississippi River Caue	0	40,341
Necedah	0	43,656
Tremplealeau	0	5,617
Upper Mississippi River	0	48,206
FH Interest	0	1,611
WYOMING		
Bamforth	0	1,166
Hutton Lake	0	1,968
Mortenson Lake	0	1,776
National Elk	0	24,774
Pathfinder	0	16,807

Unit Name	Wilderness Area Acres	Total Acres
WYOMING		
Seedskadee	0	15,723
FH Interest	0	2,000
AMERICAN SAMOA		
Rose Atoll	0	39,066
BAKER ISLAND		
Baker Island	0	31,737
JARVIS ISLAND		
Jarvis Island	0	37,519
JOHNSTON ATOLL		
Johnston Island	0	100
MIDWAY ISLANDS		
Midway Atoll	0	90,097
PUERTO RICO		
Cabo Rojo	0	587
Culebra	0	1,568
Desecheo	0	360
Laguna Cartagena	0	773
SAN JUAN ISLANDS		
Howland Island	0	32,550
VIRGIN ISLANDS		
Buck Island	0	45
Green Cay	0	14
Sandy Point	0	327
GRAND TOTAL	**20,685,372**	**88,615,896**

*Lands listed as FH Interest are managed in conjunction with the Farmer's Home Administration of the USDA.

National Wilderness Preservation System

The following table lists all wilderness areas in the United States. Administrative unit refers to the national forest, wildlife refuge, BLM district, or NPS unit in which the wilderness area is located.

Wilderness Area	Agency	Administrative Unit	Wilderness Acres
ALABAMA			
Cheaha	USFS	Talladega	7,245
Sipsey	USFS	William B. Bankhead	25,986
ALASKA			
Chuck River	USFS	Tongass	74,970
Coronation Island	USFS	Tongass	19,232
Endicott River	USFS	Tongass	98,729
Karta	USFS	Tongass	39,894
Kootznoowoo	USFS	Tongass	988,050
Kuiu	USFS	Tongass	60,581
Maurelle Islands	USFS	Tongass	4,937
Misty Fiords National Monument	USFS	Tongass	2,142,907
Petersburg Creek-Duncan Salt Chuck	USFS	Tongass	46,849
Pleasant Lemusurier-Inian	USFS	Tongass	23,151
Russell Fiord	USFS	Tongass	348,701
South Baranof	USFS	Tongass	319,568
South Etolin	USFS	Tongass	83,371
South Prince of Wales	USFS	Tongass	91,018
Stikine-Le Conte	USFS	Tongass	449,951
Tebenkof Bay	USFS	Tongass	66,839
Tracy Arm-Fords Terror	USFS	Tongass	653,179
Warren Island	USFS	Tongass	11,181
West Chichagof-Yakobi	USFS	Tongass	265,529
Aleutian Islands	USFWS	Alaska Maritime	1,300,000
Andreafsky	USFWS	Yukon Delta	1,300,000
Arctic Wildlife Refuge	USFWS	Arctic	8,000,000
Becharof	USFWS	Becharof	400,000
Bering Sea	USFWS	Alaska Maritime	81,340
Bogoslof	USFWS	Alaska Maritime	175
Chamisso	USFWS	Alaska Maritime	455
Forrester Island	USFWS	Alaska Maritime	2,832
Hazy Island	USFWS	Alaska Maritime	32
Innoko	USFWS	Innoko	1,240,000
Izembek	USFWS	Izembek	300,000
Kenai	USFWS	Kenai	1,350,000
Koyukuk	USFWS	Koyukuk	400,000

Wilderness Area	Agency	Administrative Unit	Wilderness Acres
ALASKA			
Nunivak	USFWS	Yukon Delta	600,000
Saint Lazaria	USFWS	Alaska Maritime	65
Selawik	USFWS	Selawik	240,000
Semidi	USFWS	Alaska Maritime	250,000
Simeonof	USFWS	Alaska Maritime	25,855
Togiak	USFWS	Togiak	2,270,000
Tuxedni	USFWS	Alaska Maritime	5,556
Unimak	USFWS	Alaska Maritime	910,000
Denali	NPS	National Park	2,124,783
Gates of the Arctic	NPS	NP and Preserve	7,167,192
Glacier Bay	NPS	NP and Preserve	2,664,840
Katmai	NPS	NP and Preserve	3,384,385
Kobuk Valley	NPS	National Park	174,545
Lake Clark	NPS	NP and Preserve	2,619,550
Noatak	NPS	National Preserve	5,765,427
Wrangell–St. Elias	NPS	NP and Preserve	9,078,675
ARIZONA			
Apache Creek	USFS	Prescott	5,666
Bear Wallow	USFS	Apache	11,080
Castle Creek	USFS	Prescott	25,215
Cedar Bench	USFS	Prescott	14,950
Chiricahua	USFS	Coronado	87,700
Escudilla	USFS	Apache	5,200
Fossil Springs	USFS	Coconino	22,149
Four Peaks	USFS	Tonto	61,074
Galiuro	USFS	Coronado	76,317
Granite Mountain	USFS	Prescott	9,762
Hellsgate	USFS	Tonto	37,440
Juniper Mesa	USFS	Prescott	7,406
Kachina Peaks	USFS	Coconino	18,616
Kenab Creek	USFS	Kaibab	63,760
Kendrick Mountain	USFS	Coconino, Kaibab	6,510
Mazatzal	USFS	Tonto	252,494
Miller Peak	USFS	Coronado	20,190
Mount Baldy	USFS	Apache	7,079
Mount Wrightson	USFS	Coronado	25,260
Munds Mountain	USFS	Coconino	24,411
Pajarita	USFS	Coronado	7,553
Pine Mountain	USFS	Prescott, Tonto	20,061
Pusch Ridge	USFS	Coronado	56,933
Red Rock– Secret Mountain	USFS	Coconino	47,194

Wilderness Area	Agency	Administrative Unit	Wilderness Acres
ARIZONA (Cont'd)			
Rincon Mountain	USFS	Coronado	38,590
Saddle Mountain	USFS	Kaibab	40,539
Salome	USFS	Tonto	18,531
Salt River Canyon	USFS	Tonto	32,101
Santa Teresa	USFS	Coronado	26,780
Sierra Ancha	USFS	Tonto	20,850
Strawberry Crater	USFS	Coconino	10,743
Superstition	USFS	Tonto	159,780
Sycamore Canyon	USFS	Coconino, Kaibab, Prescott	55,942
West Clear Creek	USFS	Coconino	15,238
Wet Beaver	USFS	Coconino	6,155
Woodchute	USFS	Prescott	5,833
Cabeza Prieta	USFWS	Cabeza Prieta	803,418
Havasu	USFWS	Havasu	14,606
Imperial Refuge	USFWS	Imperial	9,220
Kofa	USFWS	Kofa	516,200
Aravaipa Canyon	BLM	Safford District	19,700
Arrastra Mountain	BLM	Phoenix District	129,800
Aubrey Peak	BLM	Phoenix District	15,400
Baboquivari Peak	BLM	Safford District	2,040
Beaver Dam Mountains	BLM	Arizona Strip District	17,000
Big Horn Mountains	BLM	Phoenix District	21,000
Cottonwood Point	BLM	Arizona Strip District	6,500
Coyote Mountains	BLM	Safford District	5,100
Dos Cabezas Mountains	BLM	Safford District	11,700
Eagletail Mountains	BLM	Yuma District	100,600
East Cactus Plain	BLM	Yuma District	14,630
Fishhooks	BLM	Safford District	10,500
Gibralter Mountain	BLM	Yuma District	18,790
Grand Wash Cliffs	BLM	Arizona Strip District	36,300
Harcuvar Mountains	BLM	Yuma District	25,050
Harquahala Mountains	BLM	Phoenix District	22,880
Hassayampa River Canyon	BLM	Phoenix District	12,300
Hells Canyon	BLM	Phoenix District	10,600
Hummingbird Springs	BLM	Phoenix District	31,200
Kanab Creek	BLM	Arizona Strip District	8,850
Mount Logan	BLM	Arizona Strip District	14,600
Mount Nutt	BLM	Phoenix District	27,660

Wilderness Area	Agency	Administrative Unit	Wilderness Acres
ARIZONA			
Mount Tipton	BLM	Phoenix District	32,760
Mount Trumbull	BLM	Arizona Strip District	7,900
Mount Wilson	BLM	Phoenix District	23,900
Muggin's Mountains	BLM	Yuma District	7,640
Needle's Eye	BLM	Phoenix District	8,760
New Water Mountains	BLM	Yuma District	24,600
North Maricopa Mountains	BLM	Phoenix District	63,200
North Santa Teresa	BLM	Safford District	5,800
Paiute	BLM	Arizona Strip District	84,700
Paria Canyon- Vermillion Cliffs	BLM	Arizona Strip District	90,000
Peloncillo Mountains	BLM	Safford District	19,440
Rawhide Mountains	BLM	Yuma District	38,470
Redfield Canyon	BLM	Safford District	9,930
Sierra Estrella	BLM	Phoenix District	14,400
Signal Mountain	BLM	Phoenix District	13,350
South Maricopa Mountains	BLM	Phoenix District	60,100
Swansea	BLM	Yuma District	16,400
Table Top	BLM	Phoenix District	34,400
Tres Alamos	BLM	Phoenix District	8,300
Trigo Mountains	BLM	Yuma District	30,300
Upper Burro Creek	BLM	Phoenix District	27,440
Wabayuma Peak	BLM	Phoenix District	40,000
Warm Springs	BLM	Phoenix District	112,400
White Canyon	BLM	Phoenix District	5,790
Woolsey Peak	BLM	Phoenix District	64,000
Chiricahua	NPS	National Monument	9,440
Organ Pipe Cactus	NPS	National Monument	312,600
Petrified Forest	NPS	National Park	50,260
Saguaro	NPS	National Monument	71,400
ARKANSAS			
Black Fork Mountain	USFS	Ouachita	8,430
Caney Creek	USFS	Ouachita	14,460
Dry Creek	USFS	Ouachita	6,310
East Fork	USFS	Ozark	10,688
Flatside	USFS	Ouachita	9,507
Hurricane Creek	USFS	Ozark	15,427
Leatherwood	USFS	Ozark	16,980

Wilderness Area	Agency	Administrative Unit	Wilderness Acres
ARKANSAS			
Poteau Mountain	USFS	Ouachita	11,229
Richland Creek	USFS	Ozark	11,801
Upper Buffalo	USFS	Ozark	12,035
Big Lake	USFWS	Big Lake	2,144
Buffalo	NPS	National River	10,529
CALIFORNIA			
Agua Tibia	USFS	Cleveland	15,933
Ansel Adams	USFS	Inyo, Sierra	230,260
Bucks Lake	USFS	Plumas	21,000
Caribou	USFS	Lassen	20,546
Carson-Iceberg	USFS	Toiyabe	161,501
Castle Crags	USFS	Shasta	11,048
Chanchelulla	USFS	Trinity	8,200
Chumash	USFS	Los Padres	38,200
Cucamonga	USFS	Angeles, San Bernardino	12,781
Desolation	USFS	Eldorado	63,475
Dick Smith	USFS	Los Padres	68,000
Dinkey Lakes	USFS	Sierra	30,000
Dome Land	USFS	Sequoia	93,861
Emigrant	USFS	Stanislaus	112,338
Garcia	USFS	Los Padres	14,100
Golden Trout	USFS	Inyo, Sequoia	305,464
Granite Chief	USFS	Tahoe	25,748
Hauser	USFS	Cleveland	8,091
Hoover	USFS	Inyo, Toiyabe	48,622
Ishi	USFS	Lassen	42,866
Jennie Lakes	USFS	Sequoia	10,289
John Muir	USFS	Inyo, Sierra	581,143
Kaiser	USFS	Sierra	22,700
Machesna Mountain	USFS	Los Padres	20,000
Mantilija	USFS	Los Padres	29,600
Marble Mountain	USFS	Klamath	242,464
Mokelumne	USFS	Eldorado, Stanislaus, Toiyabe	100,848
Monarch	USFS	Sequoia, Sierra	44,896
Mount Shasta	USFS	Shasta	37,710
North Fork	USFS	Six Rivers	8,100
Pine Creek	USFS	Cleveland	13,686
Red Buttes	USFS	Rogue River	16,150
Russian	USFS	Klamath	12,000
San Gabriel	USFS	Angeles	36,118
San Gorgonio	USFS	San Bernardino	58,669
San Jacinto	USFS	San Bernardino	33,408

Wilderness Area	Agency	Administrative Unit	Wilderness Acres
CALIFORNIA (Cont'd)			
San Mateo Canyon	USFS	Cleveland	39,540
San Rafael	USFS	Los Padres	197,570
Santa Lucia	USFS	Los Padres	21,704
Santa Rosa	USFS	San Bernardino	19,803
Sespe	USFS	Los Padres	219,700
Sheep Mountain	USFS	Angeles, San Bernardino	42,367
Silver Peak	USFS	Los Padres	14,500
Siskiyou	USFS	Klamath, Siskiyou, Six Rivers	153,000
Snow Mountain	USFS	Mendocino	37,000
South Sierra	USFS	Inyo, Sequoia	60,324
South Warner	USFS	Modoc	70,729
Thousand Lakes	USFS	Lassen	16,355
Trinity Alps	USFS	Klamath, Shasta, Trinity, Six Rivers	513,100
Ventana	USFS	Los Padres	205,489
Yolla Bolly-Middle Eel	USFS	Mendocino, Trinity, Six Rivers	151,626
Havasu	USFWS	Havasu	3,195
Imperial Refuge	USFWS	Imperial Refuge	5,836
Argus Range	BLM	TBD	74,890
Big Maria Mountains	BLM	TBD	47,570
Bigelow Cholla Garden	BLM	TBD	10,380
Bighorn Mountains	BLM	TBD	39,185
Black Mountains	BLM	TBD	13,940
Bright Star	BLM	TBD	9,520
Bristol Mountains	BLM	TBD	68,515
Cadiz Dunes	BLM	TBD	39,740
Cady Mountains	BLM	TBD	84,400
Carrizo Gorge	BLM	TBD	15,700
Chemehuevi Mountains	BLM	TBD	64,320
Chimney Peak	BLM	TBD	13,700
Chuckwalla Mountains	BLM	TBD	80,770
Cleghorn Lakes	BLM	TBD	33,980
Clipper Mountains	BLM	TBD	26,000
Coso Range	BLM	TBD	50,520
Coyote Mountains	BLM	TBD	17,000
Darwin Falls	BLM	TBD	8,600
Dead Mountains	BLM	TBD	48,850

Wilderness Area	Agency	Administrative Unit	Wilderness Acres
CALIFORNIA			
Domeland Additions	BLM	TBD	36,300
El Paso Mountains	BLM	TBD	23,780
Farallon	BLM	Farallon	141
Fish Creek	BLM	TBD	25,940
Funeral Mountains	BLM	TBD	28,110
Golden Valley	BLM	TBD	37,700
Grass Valley	BLM	TBD	31,695
Hollow Hills	BLM	TBD	2,240
Ibex	BLM	TBD	26,460
Indian Pass	BLM	TBD	33,885
Inyo Mountains	BLM	TBD	205,020
Ishi	BLM	Ukiah District	240
Jacumba Mountains	BLM	TBD	33,670
Kelso Dunes	BLM	TBD	129,580
Kiavah	BLM	TBD	88,290
Kingston Range	BLM	TBD	209,608
Little Chuckwalla	BLM	TBD	29,880
Little Piacacho	BLM	TBD	33,600
Machesna Mountain	BLM	Bakersfield District	120
Malpais Mesa	BLM	TBD	32,360
Manly Peak	BLM	TBD	16,105
Mecca Hills	BLM	TBD	24,200
Mesquite	BLM	TBD	47,330
Newberry Mountains	BLM	TBD	22,900
North Algodones Dunes	BLM	TBD	32,240
North Mesquite Mountains	BLM	TBD	25,540
Old Woman Mountains	BLM	TBD	146,020
Orocopia	BLM	TBD	40,735
Owens Peak	BLM	TBD	74,640
Pahrump Valley	BLM	TBD	74,800
Palen/McCoy	BLM	TBD	220,629
Palo Verde	BLM	TBD	32,310
Picacho Peak	BLM	TBD	7,700
Piper Mountains	BLM	TBD	72,575
Piute Mountains	BLM	TBD	36,840

Wilderness Area	Agency	Administrative Unit	Wilderness Acres
CALIFORNIA			
Resting Spring Range	BLM	TBD	78,868
Rice Valley	BLM	TBD	40,820
Riverside Mountains	BLM	TBD	22,380
Rodman Mountains	BLM	TBD	21,300
Sacatar Trail	BLM	TBD	51,900
Saddle Peak Hills	BLM	TBD	1,440
San Gorgonio	BLM	TBD	37,398
Santa Lucia	BLM	Bakersfield District	1,733
Santa Rosa Mountains	BLM	TBD	64,760
Sawtooth Mountains	BLM	TBD	35,080
Sheephole Valley	BLM	TBD	174,800
South Nopah Range	BLM	TBD	16,780
Stateline	BLM	TBD	7,050
Stepladder Mountains	BLM	TBD	81,600
Surprise Canyon	BLM	TBD	29,180
Sylvania Mountains	BLM	TBD	17,820
Trilobite	BLM	TBD	31,160
Trinity Alps	BLM	Ukiah District	4,623
Turtle Mountains	BLM	TBD	144,500
Whipple Mountains	BLM	TBD	77,520
Yolla Bolly-Middle Eel	BLM	Ukiah District	8,500
Death Valley	NPS	National Park	3,158,038
Joshua Tree	NPS	National Park	561,470
Lassen Volcanic	NPS	National Park	78,982
Lava Beds	NPS	National Monument	28,460
Mojave	NPS	National Preserve	695,216
Phillip Burton	NPS	Point Reyes National Seashore	25,370
Pinnacles	NPS	National Monument	12,952
Sequoia-Kings Canyon	NPS	National Park	736,980
Yosemite	NPS	National Park	677,600

Wilderness Area	Agency	Administrative Unit	Wilderness Acres
COLORADO			
Big Blue	USFS	Uncompaghre	98,584
Cache La Poudre	USFS	Roosevelt	9,308
Collegiate Peaks	USFS	Gunnison, San Isabel, White River	167,996
Commanche Peak	USFS	Roosevelt	66,901
Eagles Nest	USFS	Arapaho, White River	133,496
Flat Tops	USFS	Routt, White River	235,230
Holy Cross	USFS	San Isabel, White River	123,410
Hunter-Fryingpan	USFS	White River	74,599
Indian Peaks	USFS	Arapahoe, Roosevelt	70,894
La Garita	USFS	Gunnison, Rio Grande	103,986
Lizard Head	USFS	San Juan, Uncompaghre	41,496
Lost Creek	USFS	Pike	105,451
Maroon Bells-Snowmass	USFS	Gunnison, White River	183,847
Mount Evans	USFS	Arapaho, Pike	74,401
Mount Massive	USFS	San Isabel	28,047
Mount Sneffels	USFS	Uncompaghre	16,587
Mount Zirkel	USFS	Routt	139,898
Neota	USFS	Roosevelt, Routt	9,924
Never Summer	USFS	Arapaho, Routt	14,100
Platte River	USFS	Routt	743
Raggeds	USFS	Gunnison, White River	59,930
Rawah	USFS	Roosevelt, Routt	73,934
South San Juan	USFS	Rio Grande, San Juan	127,690
Weminuche	USFS	Rio Grande, San Juan	463,678
West Elk	USFS	Gunnison	176,412
Leadville	USFWS	Leadville National Fish Hatchery	2,560
Black Canyon of the Gunnison	NPS	National Monument	11,180
Great Sand Dunes	NPS	National Monument	33,450
Mesa Verde	NPS	National Park	8,100
FLORIDA			
Alexander Springs	USFS	Ocala	7,941
Big Gum Swamp	USFS	Osceola	13,660
Billies Bay	USFS	Ocala	3,092
Bradwell Bay	USFS	Apalachicola	24,602
Juniper Prairies	USFS	Ocala	14,281
Little Lake George	USFS	Ocala	2,833
Mud Swamp/ New River	USFS	Apalachicola	8,090

Wilderness Area	Agency	Administrative Unit	Wilderness Acres
FLORIDA			
Cedar Keys	USFWS	Cedar Keys	397
Chassahowitzka	USFWS	Chassahowitzka	23,580
Florida Keys	USFWS	Great White Huron	1,900
Florida Keys	USFWS	Key West	2,019
Florida Keys Deer Refuge	USFWS	National Key	2,278
Island Bay	USFWS	Island Bay	20
J. N. Ding Darling	USFWS	J. N. Ding Darling	2,619
Lake Woodruff	USFWS	Lake Woodruff	1,066
Passage Key	USFWS	Passage Key	36
Pelican Island	USFWS	Pelican Island	6
Saint Marks	USFWS	Saint Marks	17,350
Everglades	NPS	National Park	1,296,500
GEORGIA			
Big Frog	USFS	Chattahoochee	83
Blood Mountain	USFS	Chattahoochee	7,800
Brasstown	USFS	Chattahoochee	12,565
Cohutta	USFS	Chattahoochee	35,147
Ellicott Rock	USFS	Chattahoochee	2,181
Mark Trail	USFS	Chattahoochee	16,400
Raven Cliffs	USFS	Chattahoochee	8,562
Rich Mountain	USFS	Chattahoochee	9,649
Southern Nantahala	USFS	Chattahoochee	11,770
Tray Mountain	USFS	Chattahoochee	9,702
Blackbeard Island	USFWS	Blackbeard Island	3,000
Okefenokee	USFWS	Okefenokee	353,981
Wolf Island	USFWS	Wolf Island	5,126
Cumberland Island	NPS	National Seashore	8,840
HAWAII			
Haleakala	NPS	National Park	19,270
Hawaii Volcanoes	NPS	National Park	123,100
IDAHO			
Frank Church- River of No Return	USFS	Bitterroot, Boise, Challis, Nez Perce, Fayette, Salmon	2,361,767
Gospel Hump	USFS	Nez Perce	206,000
Hells Canyon	USFS	Nez Perce, Payette	84,100
Sawtooth	USFS	Boise, Challis, Sawtooth	217,088
Selway-Bitterrot	USFS	Bitterroot, Clearwater, Nez Perce	1,089,238
Frank Church- River of No Return	BLM	Boise National Forest	720
Craters of the Moon	NPS	National Monument	43,243

Wilderness Area	Agency	Administrative Unit	Wilderness Acres
ILLINOIS			
Bald Knob	USFS	Shawnee	5,918
Bay Creek	USFS	Shawnee	2,866
Burden Falls	USFS	Shawnee	3,723
Clear Springs	USFS	Shawnee	4,730
Garden of the Gods	USFS	Shawnee	3,293
Lusk Creek	USFS	Shawnee	4,796
Panther Den	USFS	Shawnee	940
Crab Orchard	USFWS	Crab Orchard	4,050
INDIANA			
Charles C. Deam	USFS	Hoosier	12,953
KENTUCKY			
Beaver Creek	USFS	Daniel Boone	4,791
Clifty	USFS	Daniel Boone	12,646
LOUISIANA			
Kisatchie Hills	USFS	Kisatchie	8,679
Breton	USFWS	Breton	5,000
Lacassine	USFWS	Lacassine	3,346
MAINE			
Caribou- Speckled Mtn.	USFS	White Mountain	12,000
Moosehorn, Baring Unit	USFWS	Moosehorn	4,680
Moosehorn, Birch Islands Unit	USFWS	Moosehorn	6
Moosehorn, Edmunds Unit	USFWS	Moosehorn	2,706
MASSACHUSETTS			
Monomoy	USFWS	Monomoy	2,420
MICHIGAN			
Big Island Lake	USFS	Hiawatha	6,008
Delirium	USFS	Hiawatha	12,000
Horseshoe Bay	USFS	Hiawatha	3,949
Mackinac	USFS	Hiawatha	12,388
McCormick	USFS	Ottawa	16,850
Nordhouse Dunes	USFS	Manistee	3,450
Rock River Canyon	USFS	Hiawatha	5,285
Round Island	USFS	Hiawatha	378
Sturgeon River Gorge	USFS	Ottawa	14,850

Wilderness Area	Agency	Administrative Unit	Wilderness Acres
MICHIGAN			
Sylvania	USFS	Ottawa	18,327
Huron Islands	USFWS	Huron	147
Michigan Islands	USFWS	Michigan Islands	12
Seney	USFWS	Seney	25,150
Isle Royale	NPS	National Park	131,880
MINNESOTA			
Boundary Waters Canoe Area	USFS	Superior	1,086,954
Agassiz	USFWS	Agassiz	4,000
Tamarac	USFWS	Tamarac	2,180
MISSISSIPPI			
Black Creek	USFS	DeSoto	5,052
Leaf	USFS	DeSoto	994
Gulf Islands	NPS	National Seashore	3,202
MISSOURI			
Bell Mountain	USFS	Mark Twain	9,027
Devil's Backbone	USFS	Mark Twain	6,595
Hercules-Glades	USFS	Mark Twain	12,315
Irish	USFS	Mark Twain	16,358
Paddy Creek	USFS	Mark Twain	7,059
Piney Creek	USFS	Mark Twain	8,142
Rockpile Mountain	USFS	Mark Twain	4,131
Mingo	USFWS	Mingo	7,730
MONTANA			
Absaroka-Beartooth	USFS	Custer, Gallatin	920,310
Anaconda-Pintler	USFS	Beaverhead, Bitterroot, Deerlodge	157,874
Bob Marshall	USFS	Flathead, Lewis and Clark	1,009,356
Cabinet Mountains	USFS	Kaniksu, Kootenai	94,272
Gates of the Mountains	USFS	Helena	28,562
Great Bear	USFS	Flathead	286,700
Lee Metcalf	USFS	Beaverhead, Gallatin	248,944
Mission Mountains	USFS	Flathead	73,877
Rattlesnake	USFS	Lolo	33,000
Scapegoat	USFS	Helena, Lewis and Clark, Lolo	239,296
Selway-Bitterroot	USFS	Bitterroot, Lolo	251,343
Welcome Creek	USFS	Lolo	28,135
Medicine Lake	USFWS	Medicine Lake	11,366
Red Rock Lakes	USFWS	Red Rock Lakes	32,350
UL Bend	USFWS	UL Bend	20,819

Wilderness Area	Agency	Administrative Unit	Wilderness Acres
MONTANA			
Lee Metcalf-Bear Trap Canyon Unit	BLM	Butte District	6,000
NEBRASKA			
Soldier Creek	USFS	Nebraska	7,794
Fort Niobrara	USFWS	Fort Niobrara	4,635
NEVADA			
Alta Toquima	USFS	Toiyabe	38,000
Arc Dome	USFS	Toiyabe	115,000
Boundary Peak	USFS	Inyo	10,000
Currant Mountain	USFS	Humboldt	36,000
East Humboldts	USFS	Humboldt	36,900
Grant Range	USFS	Humboldt	50,000
Jarbidge	USFS	Humboldt	113,327
Mount Charlston	USFS	Toiyobe	43,000
Mount Rose	USFS	Toiyobe	28,000
Mount Moriah	USFS	Humboldt	70,000
Quinn Canyon	USFS	Humboldt	27,000
Ruby Mountains	USFS	Humboldt	90,000
Santa Rosa-Paradise Peak	USFS	Humboldt	31,000
Table Mountain	USFS	Toiyabe	98,000
Arc Dome	BLM	Battle Mountain District	20
Currant Mountain	BLM	Ely District	3
Mount Moriah	BLM	Ely District	6,435
NEW HAMPSHIRE			
Great Gulf	USFS	White Mountain	5,552
Pemigewasset	USFS	White Mountain	45,000
Presidential Range-Dry River	USFS	White Mountain	27,380
Sandwich Range	USFS	White Mountain	25,000
NEW JERSEY			
Brigantine	USFWS	Edwin B. Forsythe	6,681
Great Swamp	USFWS	Great Swamp	3,660
NEW MEXICO			
Aldo Leopold	USFS	Gila	202,016
Apache Kid	USFS	Cibola	44,626
Blue Range	USFS	Apache, Gila	30,000
Captain Mountains	USFS	Lincoln	35,967
Chama River Canyon	USFS	Carson, Santa Fe	50,300
Cruces Basin	USFS	Carson	18,000

Wilderness Area	Agency	Administrative Unit	Wilderness Acres
NEW MEXICO			
Dome	USFS	Santa Fe	5,200
Gila	USFS	Gila	558,065
Latir Peak	USFS	Carson	20,000
Manzano Mountain	USFS	Cibola	37,195
Pecos	USFS	Carson, Santa Fe	223,333
San Pedro Parks	USFS	Santa Fe	41,132
Sandia Mountain	USFS	Cibola	38,357
Wheeler Peak	USFS	Carson	19,661
White Mountain	USFS	Lincoln	48,885
Withington	USFS	Cibola	19,000
Bosque del Apache, Chupadera Unit	USFWS	Bosque del Apache	5,289
Bosque del Apache, Indian Well Unit	USFWS	Bosque del Apache	5,139
Bosque del Apache, San Pasqual Unit	USFWS	Bosque del Apache	19,859
Salt Creek	USFWS	Bitter Lake	9,621
Bisti	BLM	Albuquerque District	3,968
Cebolla	BLM	Albuquerque District	62,800
De-na-zin	BLM	Albuquerque District	23,872
West Malpais	BLM	Albuquerque District	39,700
Bandelier	NPS	National Monument	23,267
Carlsbad Caverns	NPS	National Park	33,125
NEW YORK			
Fire Island	NPS	National Seashore	1,363
NORTH CAROLINA			
Birkhead Mountains	USFS	Uwharrie	5,160
Catfish Lake South	USFS	Croatan	8,530
Ellicott Rock	USFS	Nantahala	4,022
Joyce Kilmer-Slickrock	USFS	Nantahala	13,562
Linville Gorge	USFS	Pisgah	12,002
Middle Prong	USFS	Pisgah	7,460
Pocosin	USFS	Croatan	11,709
Pond Pine	USFS	Croatan	1,685
Sheep Ridge	USFS	Croatan	9,297
Shining Rock	USFS	Pisgah	18,483
Southern Nantahala	USFS	Nantahala	11,944
NORTH DAKOTA			
Chase Lake	USFWS	Chase Lake	4,155
Lostwood	USFWS	Lostwood	5,577
Theodore Roosevelt	NPS	National Park	29,920

Wilderness Area	Agency	Administrative Unit	Wilderness Acres
OHIO			
West Sister Island	USFWS	West Sister Island	77
OKLAHOMA			
Black Fork Mountain	USFS	Ouachita	5,149
Upper Kiamichi	USFS	Ouachita	10,819
Wichita Mountains, Charons Garden Unit	USFWS	Wichita Mountains	5,723
Wichita Mountains, North Mountain Unit	USFWS	Wichita Mountains	2,847
OREGON			
Badger Creek	USFS	Mt. Hood	24,000
Black Canyon	USFS	Ochoco	13,400
Boulder Creek	USFS	Umpqua	19,100
Bridge Creek	USFS	Ochoco	5,400
Bull of the Woods	USFS	Mt. Hood, Willamette	34,900
Columbia	USFS	Mt. Hood	39,000
Cummins Creek	USFS	Siuslaw	9,173
Diamond Peak	USFS	Deschutes, Willamette	52,337
Drift Creek	USFS	Suislaw	5,798
Eagle Cap	USFS	Wallowa, Whitman	358,461
Gearhart Mountain	USFS	Fremont	22,809
Grassy Knob	USFS	Siskiyou	17,200
Hells Canyon	USFS	Wallowa, Whitman	130,095
Kalmiopsis	USFS	Siskiyou	179,700
Menagerie	USFS	Willamette	4,800
Middle Santiam	USFS	Willamette	7,500
Mill Creek	USFS	Ochoco	17,400
Monument Rock	USFS	Malheur, Whitman	19,800
Mount Hood	USFS	Mount Hood	47,160
Mount Jefferson	USFS	Deschutes, Mt. Hood, Willamette	107,008
Mount Theilsen	USFS	Umpqua, Willamette, Winema	54,267
Mount Washington	USFS	Deschutes, Willamette	52,738
Mountain Lake	USFS	Winema	23,071
North Fork John Day	USFS	Umatilla, Whitman	121,400
North Fork Umatilla	USFS	Umatilla	20,435
Red Buttes	USFS	Siskiyou, Rogue River	3,750
Rock Creek	USFS	Siuslaw	7,486
Rogue-Umpqua Divide	USFS	Rogue River, Umpqua	33,200
Salmon-Huckleberry	USFS	Mt. Hood	44,600
Sky Lakes	USFS	Rogue River, Winema	116,300
Strawberry Mountain	USFS	Malheur	69,350

Wilderness Area	Agency	Administrative Unit	Wilderness Acres
OREGON			
Three Sisters	USFS	Deschutes, Willamette	285,202
Waldo Lake	USFS	Willamette	39,200
Wenaha-Tucannon	USFS	Umatilla	66,417
Wild Rogue	USFS	Siskiyou	36,500
Oregon Islands	USFWS	Oregon Islands	480
Three Arch Rocks	USFWS	Three Arch Rocks	15
Hells Canyon	BLM	Vale District	1,038
Oregon Islands	BLM	Coos Bay District	5
Table Rock	BLM	Salem District	5,500
Wild Rogue	BLM	Medford District	10,000
PENNSYLVANIA			
Allegheny Islands	USFS	Allegheny	368
Hickory Creek	USFS	Allegheny	8,570
SOUTH CAROLINA			
Ellicott Rock	USFS	Sumter	2,809
Hell Hole Bay	USFS	Francis Marion	2,180
Little Wambaw Swamp	USFS	Francis Marion	5,154
Wambaw Creek	USFS	Francis Marion	1,937
Wambaw Swamp	USFS	Francis Marion	4,767
Cape Romain	USFWS	Cape Romain	29,000
Congaree Swamp	NPS	National Monument	15,000
SOUTH DAKOTA			
Black Elk	USFS	Black Hills	9,826
Badlands	NPS	National Park	64,250
TENNESSEE			
Bald River Gorge	USFS	Cherokee	3,721
Big Frog	USFS	Cherokee	7,986
Big Laurel Branch	USFS	Cherokee	6,251
Citico Creek	USFS	Cherokee	16,226
Cohutta	USFS	Cherokee	1,795
Gee Creek	USFS	Cherokee	2,493
Joyce Kilmer-Slickrock	USFS	Cherokee	3,832
Little Frog Mountain	USFS	Cherokee	4,684
Pond Mountain	USFS	Cherokee	6,665
Sampson Mountain	USFS	Cherokee	7,992
Unaka Mountain	USFS	Cherokee	4,700
TEXAS			
Big Slough	USFS	Davey Crockett	3,455
Indian Mounds	USFS	Sabine	10,917

Wilderness Area	Agency	Administrative Unit	Wilderness Acres
TEXAS			
Little Lake Creek	USFS	Sam Houston	38,553
Turkey Hill	USFS	Angelina	5,473
Upland Island	USFS	Angelina	12,650
Guadalupe Mountains	NPS	National Park	46,850
UTAH			
Ashdown Gorge	USFS	Dixie	7,000
Box-Death Hollow	USFS	Dixie	25,814
Dark Canyon	USFS	Manti-LeSal	45,000
Deseret Peak	USFS	Wasatch	25,500
High Uintas	USFS	Ashley, Wasatch	460,000
Lone Peak	USFS	Uinta, Wasatch	30,088
Mount Naomi	USFS	Cache	44,350
Mount Nebo	USFS	Uinta	28,000
Mount Olympus	USFS	Wasatch	16,000
Mount Timpanogos	USFS	Uinta	10,750
Pine Valley Mountain	USFS	Dixie	50,000
Twin Peaks	USFS	Wasatch	11,463
Wellsville Mountain	USFS	Cache	23,850
Beaver Dam Mountains	BLM	Cedar City Districts	2,600
Paria Canyon-Vermillion Cliffs	BLM	Cedar City Districts	20,000
VERMONT			
Big Branch	USFS	Green Mountain	6,720
Breadloaf	USFS	Green Mountain	21,480
Bristol Cliffs	USFS	Green Mountain	3,738
George D. Aiken	USFS	Green Mountain	5,060
Lye Brook	USFS	Green Mountain	15,680
Peru Peak	USFS	Green Mountain	6,920
VIRGINIA			
Barbours Creek	USFS	George Washington, Jefferson	5,700
Beartown	USFS	Jefferson	5,609
James River Face	USFS	Jefferson	8,886
Kimberling Creek	USFS	Jefferson	5,542
Lewis Fork	USFS	Jefferson	5,618
Little Dry Run	USFS	Jefferson	2,858
Little Wilson Creek	USFS	Jefferson	3,613
Mountain Lake	USFS	Jefferson	8,392
Peters Mountain	USFS	Jefferson	3,328
Ramseys Draft	USFS	George Washington	6,518

Wilderness Area	Agency	Administrative Unit	Wilderness Acres
VIRGINIA			
Rich Hole	USFS	George Washington	6,450
Rough Mountain	USFS	George Washington	9,300
Saint Marys	USFS	George Washington	9,835
Shawyers Run	USFS	George Washington, Jefferson	3,665
Thunder Ridge	USFS	Jefferson	2,344
Shenandoah	NPS	National Park	79,579
WASHINGTON			
Alpine Lakes	USFS	Snoqualmie, Wenatchee	364,229
Boulder River	USFS	Mount Baker	48,674
The Brothers	USFS	Olympic	16,682
Buckhorn	USFS	Olympic	44,474
Clearwater	USFS	Snoqualmie	14,374
Colonel Bob	USFS	Olympic	11,961
Glacier Peak	USFS	Mount Baker, Wenatchee	572,738
Glacier View	USFS	Gifford Pinchot	3,123
Goat Rocks	USFS	Gifford Pinchot, Snoqualmie	108,439
Henry M. Jackson	USFS	Mount Baker, Snoqualmie, Wenatchee	103,593
Indian Heaven	USFS	Gifford Pinchot	20,960
Lake Chelan-Sawtooth	USFS	Okanogan, Wenatchee	151,564
Mount Adams	USFS	Gifford Pinchot	56,861
Mount Baker	USFS	Mount Baker	117,848
Mount Skokamish	USFS	Olympic	13,015
Noisy-Diobsud	USFS	Mount Baker	14,133
Norse Peak	USFS	Snoqualmie, Wenatchee	50,180
Pasayten	USFS	Mount Baker, Okangogan	530,031
Salmo Priest	USFS	Colville, Kaniksu	41,335
Tatoosh	USFS	Gifford Pinchot	15,750
Trapper Creek	USFS	Gifford Pinchot	5,970
Wenaha-Tucannon	USFS	Umatilla	111,048
William O. Douglas	USFS	Gifford Pinchot, Wenatchee	168,157
Wonder Mountain	USFS	Olympic	2,349
San Juan Islands	USFWS	San Juan Islands	353
Washington Islands	USFWS	Copalis	60
Washington Islands	USFWS	Flattery Rocks	125
Washington Islands	USFWS	Quallayute Needles	300
Juniper Dunes	BLM	Spokane District	7,140
Mount Rainier	NPS	National Park	216,855
North Cascades	NPS	National Park	634,614
Olympic National Park	NPS	National Park	876,669

Wilderness Area	Agency	Administrative Unit	Wilderness Acres
WEST VIRGINIA			
Cranberry	USFS	Monongahela	35,864
Dolly Sods	USFS	Monongahela	10,215
Laurel Fork North	USFS	Monongahela	6,055
Laurel Fork South	USFS	Monongahela	5,997
Mountain Lake	USFS	Jefferson	2,721
Otter Creek	USFS	Monongahela	20,000
WISCONSIN			
Blackjack Springs	USFS	Nicolet	5,886
Headwaters	USFS	Nicolet	20,104
Porcupine Lake	USFS	Chequamegon	4,446
Rainbow Lake	USFS	Chequamegon	6,583
Whisker Lake	USFS	Nicolet	7,428
Wisconsin Islands	USFWS	Gravel Island	27
Wisconsin Islands	USFWS	Green Bay	2
WYOMING			
Absaroka-Beartooth	USFS	Shoshone	23,283
Bridger	USFS	Bridger	428,087
Cloud Peak	USFS	Bighorn	189,039
Encampment River	USFS	Medicine Bow	10,124
Fitzpatrick	USFS	Shoshone	198,525
Gros Ventre	USFS	Bridger, Teton	287,000
Huston Park	USFS	Medicine Bow	30,726
Jedediah Smith	USFS	Targhee	123,451
North Absaroka	USFS	Shoshone	350,488
Platte River	USFS	Medicine Bow	22,749
Popo Agie	USFS	Shoshone	101,870
Savage Run	USFS	Medicine Bow	14,930
Teton	USFS	Teton	585,238
Washakie	USFS	Shoshone	704,822
Winegar Hole	USFS	Targhee	10,715

Forest Service Total	**34,469,854**
Fish and Wildlife Service Total	**20,685,372**
Bureau of Land Management Total	**5,176,165**
Park Service Total	**43,107,581**
GRAND TOTAL	**103,438,972**

Source: The information in this table was compiled from the individual agencies responsible for wilderness area administration: the United States Department of Agriculture's Forest Service, and the Department of the Interior's Bureau of Land Management, Fish and Wildlife Service, and National Park Service.

6

Directory of Federal Agencies, Private Organizations, and Associations

THE FOLLOWING DIRECTORY INCLUDES the federal agencies that are responsible for setting and administering U.S. policy relating to the American wilderness, and private organizations and associations in which wildlife and wilderness are the major concern.

Federal Agencies

Environmental Protection Agency (EPA)
401 M Street, SW
Washington, DC 20024
(202) 260-2090

The EPA is the largest independent regulatory agency in the federal government. It is responsible for enforcement of environmental laws on issues of clean water, air, solid waste disposal, pesticide research and registration, and radiation monitoring.

United States Army Corps of Engineers
U.S. Department of Defense
Office of the Chief of Engineers
Pulaski Building
20 Massachusetts Ave., NW
Washington, DC 20314
(202) 272-0001

The Army Corps of Engineers is responsible for any work within navigable waters. Their projects can include dams, reservoirs, and flood control
levees, and they are often involved in activities on or around wetlands.

United States Department of the Interior
1849 C Street, NW
Washington, DC 20240
(202) 208-3100

The Department of the Interior was formed in 1849 as a cabinet level
department of the executive branch. The department is headed by the
Secretary of the Interior, who is appointed by and reports directly to the
President of the United States.

The Interior Department was originally formed in order to organize
a number of U.S. agencies under one roof. These included the Bureau
of Indian Affairs, the General Land Office, Patent Office, and Pension
Office. More offices were added to the department over time and it began
to focus on management of the nation's vast natural resources. Today the
department has over 30 different offices and bureaus, including

> **U.S. Geological Survey:** Responsible for mapping and surveying
> the lands within the United States.
>
> **Fish and Wildlife Service (USFWS):** Oversees lands that are to
> be managed primarily for the benefit of native fish and wildlife
> and also manages all national wildlife refuges, National Fish
> Hatcheries, Fisheries Research Centers, Migratory Waterfowl
> Refuges, and Waterfowl Production Areas.
>
> **Bureau of Indian Affairs:** Responsible for all federal
> conservation programs on Indian lands.
>
> **Bureau of Land Management (BLM):** Manages all public domain
> lands that have not been set aside for a specific purpose, such as
> the national parks or forests.
>
> **Bureau of Mines:** Oversees all mining operations on federal
> lands.
>
> **National Park Service (NPS):** Manages the National Park System.

Bureau of Reclamation: Responsible for water-related activities on non-navigable waters, such as ground water recharging, construction of water systems, and building major dams.

The goals of the Interior Department have evolved since its inception. Originally the country's resources were seen as inexhaustible, and the department's job was to dole out the rights to use those resources in order to best help the economy. Today we see that our resources are limited, and while it is still the department's job to allocate rights to developers, this is tempered by the ideas of management, conservation, and preservation, to ensure that we still have these resources in the future and to save significant natural areas for posterity. The degree to which the Interior Department favors preservation or development depends to a large degree on the views of the acting secretary and presidential administration.

United States Forest Service (USFS)
United States Department of Agriculture (USDA)
Washington, DC 20250
(202) 205-1760

The USFS is responsible for managing the national forests, regulating logging, grazing, mining, recreation, wilderness preservation, and any other activities that take place within these forests.

Private Organizations and Associations

Alaska Conservation Foundation (ACF)
430 West 7th Street, Suite 215
Anchorage, AK 99501
(907) 276-1917
FAX (907) 274-4145
Founded: 1961

A fund-raising organization that distributes grants to Alaskan environmental groups. The ACF has distributed grants totaling more than $2 million to 74 different organizations which help protect ocean habitats, wildlife, and wildlands. After the Exxon Valdez oil spill in 1989, the ACF immediately created the Prince William Sound Clean Up and Rehabilitation Fund in cooperation with Governor Cowper to manage the large scale of donations flooding into the state.

PUBLICATIONS: *ACF Annual Report* details ACF's projects, *Issue Updates* is published periodically and addresses current Alaskan environmental issues, and a *Grantseekers Guide* explains how to apply for a grant.

American Forest Council
1250 Connecticut Ave., NW
Washington, DC 20036
(202) 463-2455
Founded: 1941

The American Forest Council is an organization representing the U.S. forest products industry. The council promotes logging and commercial development of the nation's public and private forests.

American Forestry Association
1561 P Street NW
Washington, DC 20005
(202) 667-3300
Founded: 1875

This is primarily a lobbying group concerned with forest conservation. Their program best known to the general public is called *Global ReLeaf*, and is geared toward planting trees and encouraging others to do the same around the world.

PUBLICATIONS: *American Forests* is a bimonthly magazine, *Urban Forest FORUM* is a bimonthly newsletter, and *Global ReLeaf Report* is a quarterly newsletter.

American Hiking Society
P.O. Box 20160, NW
Washington, DC 20041
(703) 385-3252

This society, with 100+ affiliates throughout the country, raises funds for construction and maintenance of trails, provides public information, and works to protect wild hiking areas. They organize National Trails Day, and are currently working on a project to create the American Discovery Trail, which will wind 5,500 miles from coast to coast.

PUBLICATIONS: *American Hiker* is a quarterly magazine for members. *Pathways Across America* is a directory of trails. *Helping Out in the Outdoors* is a listing of more than 2,000 volunteer jobs in the wilderness.

American Rivers Inc.
801 Pennsylvania Avenue, SE, Suite 400
Washington DC 20003
(202) 547-6900
(800) 783-2199
FAX (202) 543-6142
Founded: 1973

American Rivers works to preserve America's rivers and protect them from overdevelopment and hydroelectric dams. American Rivers acknowledges hydroelectric power as a viable energy source, yet endeavors to find a balance between energy and environmental needs. American Rivers also works to add rivers to the Wild and Scenic Rivers System, and has helped preserve over 9,000 miles of rivers and save 7 million acres of adjacent lands from flooding.

American Rivers monitors all forest plans from the BLM, the USFS, the Federal Energy Regulatory Commission, and the NPS. They lobby to prevent or limit dams before they are built, or to have them torn down when their licenses have expired.

PUBLICATIONS: *American Rivers* is a 24-page quarterly newsletter. The organization has also published three books, *American Rivers Guide to Wild and Scenic River Designation, American Rivers Outstanding Rivers List,* and *Rivers at Risk: The Concerned Citizen's Guide to Hydropower.*

American Wildlands (AW)
7500 East Arapaho Road, Suite 355
Englewood, CO 80112
(303) 771-0380
Founded: 1986

Formerly called American Wilderness Alliance, and formed to manage and protect America's publicly owned wildlands and monitor timber management policies. In 1986 AW stopped the sale of 4.2 million board feet of timber near the Electric Peak Wilderness Area in Yellowstone Park. AW has also been involved in:

Elephant research in cooperation with Dr. Richard Leakey, Director of Wildlife Services in Kenya.

Protecting ancient juniper forests in the Carrizo mountains, New Mexico.

Preventing oil drilling in the Arctic National Wildlife refuge.

Lobbying Congress to adopt a river legislation package.

American Wildlands also runs a program called Recreation-Conservation Connection to create an interest in outdoor recreation activities that do not damage the environment. AW offers more than 80 adventure trips each year to places such as Antarctica, Rwanda, Belize, the former Soviet Union, and throughout the United States.

PUBLICATIONS: *Wild America* is an annual magazine. *On the Wild Side* is a quarterly newsletter.

Center for Conservation Biology
Department of Biological Sciences
Stanford University
Stanford, CA 94305
(415) 723-5924
FAX (415) 723-5920
Founded: 1984

This is a scientific research organization for scientists working on conservation projects relating to all types of biological environments.

Center for Plant Conservation
125 Arborway
Jamaica Plain, MA 02130
(617) 524-6988
Founded: 1984

In order to protect endangered plant species native to the United States, the Center for Plant Conservation created the National Collection of Endangered Plants. Specimens of endangered plants are distributed to botanic gardens across the United States for protection and study. Seeds are also stored in a seed bank managed by the National Plant Germplasm System of the USDA. So far more than 300 species of plants have entered the collection. The center also works with the BLM to protect, study, and reintroduce rare and endangered plants on government lands, and to train BLM employees in plant conservation methods.

PUBLICATION: *Plant Conservation,* a quarterly report.

Conservation Law Foundation of New England
3 Joy Street
Boston, MA 02108-1497
(617) 742-2540
Founded: 1966

The Conservation Law Foundation is a group of lawyers and scientists that work to protect the environment of New England from waste and pollution through education and litigation. They also work with local utility companies to develop energy efficiency programs.

Some of the Conservation Law Foundation's achievements include successfully blocking the U.S. government from oil and gas drilling on George's Bank, the most productive fishing area in the world; launching cleanup programs for several harbors, including Boston Harbor; and forcing the Massachusetts Military Reservation to stop several polluting procedures.

PUBLICATIONS: The organization publishes the *Conservation Law Foundation of New England Quarterly Newsletter.*

Defenders of Wildlife
1244 Nineteenth Street, NW
Washington, DC 20036
(202) 659-9510
Founded: 1947

Defenders of Wildlife was founded to protect plants, animals, and animal habitats in the United States. The organization has successfully lobbied Congress to set aside millions of dollars in funds for specific projects, including $1 million to expand the Rachel Carson National Wildlife Refuge in Maine, $2 million for the Sacramento River National Wildlife Refuge in California, and $10 million for an expansion of the Lower Rio Grande National Wildlife Refuge in Texas. The Defenders have also worked to keep cattle from grazing on ecologically sensitive land administered by the BLM. Currently the organization has more than 80,000 members.

PUBLICATIONS: *Defenders* is a bimonthly magazine. *The Activists Newsletter* is for volunteers in the Defenders Activist Network, which includes 9,000 members.

Earth First!
P.O. Box 5871
Tucson, AZ 85703
(602) 622-1371
Founded: 1980

As an environmental group, Earth First! stands alone in its organization and philosophies. Earth First! is not an association or organization, but considers itself a movement. No dues are required, but those who want to be a part of the movement can purchase a subscription to an Earth First! newsletter, which provides valuable conservation information and updates Earth First!ers on current campaigns.

Unlike traditional conservation organizations, Earth First! believes that it takes more to change environmental policy than just oral or written arguments. Many consider that a legitimate way to fight back against those who destroy the environment is through physical means, such as chaining themselves to tree crushers, driving spikes into trees so that they cannot be cut, and damaging bulldozers and other heavy equipment. This ideology makes Earth First! one of the most controversial environmental groups—even among environmentalists.

The cofounder and driving force behind Earth First! for ten years, Dave Foreman, left in 1990 amid mounting pressure and FBI investigations. Today the Earth First! movement continues to organize protests and rally for conservation in local chapters across the country and around the world.

PUBLICATIONS: *Earth First!, The Radical Environmental Journal* is published eight times per year. The journal is the main link between the local Earth First! chapters, highlighting current projects, activities, and future plans of the movement.

Earth Island Institute
300 Broadway, Suite 28
San Francisco, CA 94133
(415) 788-3666
FAX (415) 788-7324
Telex: 6502829302 MCI UW
Founded: 1982

Earth Island Institute is a globally oriented environmental organization with more than 30,000 members organized through a national headquarters and local Earth Island Centers. The institute directs specific projects to aid the environment, including

> Conferences on the Fate of the Earth: Held every two years, these international conferences address the state of the earth's environment and related economic and political topics.

> Friends of the Ancient Forests: Works to preserve old growth forests of the Pacific Northwest by informing the public.

> Green Alternative Information for Action: Helps green organizations in the United States communicate with each other in order to work together better as a unified movement.

The institute also works to protect marine mammals and was the first to film the mass slaughter of dolphins by tuna fisher's driftnets. Earth Island Institute's founder, David R. Brower, also founded Friends of the Earth, and he was the first executive director of the Sierra Club.

PUBLICATIONS: *The Earth Island Journal* is a quarterly magazine for members. The institute also releases reports and publishes the *Green Calendar for Social Change.*

Earthwatch
680 Mount Auburn Street
P.O. Box 403
Watertown, MA 02272-9104
(617) 926-8200
Founded: 1971

Since 1971 Earthwatch has sent volunteers around the world to work with scientists in an effort to protect wildlife habitats, endangered species and rainforests, and to solve environmental problems. Volunteers pay a share of the cost of their trip, which can range from $700 to $2,500. These expeditions go anywhere from the rainforests of Australia to the Shenandoah valley in Virginia, or to the plains of Namibia. The goal of these trips is to study specific habitats around the world and to increase public awareness. A briefing is written for each expedition, including the history, background, and mission of the project, the background of the scientists and volunteers involved, maps and other information on the region, and a bibliography.

PUBLICATIONS: *Earthwatch Magazine* is published six times per year for members. Expedition briefings are also available to all members for a fee.

Environmental Defense Fund (EDF)
257 Park Avenue South
New York, NY 10010
(212) 505-2100
Founded: 1967

The Environmental Defense Fund (EDF) began as an organization with the primary tactic of raising money to litigate against companies that did damage to the environment. Today, the EDF continues with its legal battles, but has broadened its activities to be more of a mainstream environmental organization that supports projects aimed at research and solutions to problems of global warming, rain forest destruction, toxic wastes, acid rain, and recycling.

PUBLICATIONS: The *EDF Letter* is a bimonthly newsletter for members.

Friends of the Earth (FOE)
218 D Street, SE
Washington, DC 20003
(202) 544-2600
Founded: 1969

Friends of the Earth (FOE) was founded by David Brower, who was also responsible for starting Earth Island Institute. Friends of the Earth currently has 50,000 members with offices in Washington D.C., Seattle, and Manila, the Philippines.

FOE has seven major areas of concern, which include ozone depletion, tropical rainforest destruction, global warming, waste disposal, nuclear weapons production, corporate accountability, and oceans and coasts. Along with other environmental groups, Friends of the Earth has helped to protect 103 million acres in the Arctic National Wildlife Refuge.

PUBLICATIONS: Ten times a year FOE publishes a newsmagazine for members titled *Not Man Apart,* and several quarterly newsletters, including *Atmosphere,* on ozone depletion, *Community Plume,* on chemical safety, and *Groundwater News.*

Grand Canyon Trust
The Homestead
Route 4, Box 718
Flagstaff, AZ 86001
(602) 774-7488
1400 Sixteenth Street, Suite 300
Washington, DC 20036
(202) 797-5429
Founded: 1985

The Grand Canyon Trust works to preserve the Grand Canyon, Glen Canyon, and surrounding wilderness areas. Their major concern is air pollution within the Grand Canyon.

PUBLICATION: *Colorado Plateau Advocate* is a quarterly newsletter.

Greater Ecosystem Alliance
P.O. Box 2813
Bellingham, WA 98227
(206) 671-9950
Founded: 1989

This organization concentrates on protecting forest ecosystems in Washington State, particularly on the Olympic Peninsula and within the Selkirk and Cascade ranges.

PUBLICATION: *Northwest Conservation* is a quarterly newsletter.

Greater Yellowstone Coalition
P.O. Box 1874
Bozeman, MT 59771
(406) 586-1593
Founded: 1983

The Greater Yellowstone Coalition was formed to protect the Greater Yellowstone ecosystem, which includes Yellowstone and Grand Teton National Parks, seven national forests, three national wildlife refuges, BLM lands, and over 1 million acres of private land, for a total of 14 million acres of wilderness. The goal of the coalition is to have the natural boundaries of the ecosystem recognized. These boundaries stretch farther than official park boundaries, but the coalition holds that the extra areas are an essential part of the plant and animal habitat.

The coalition collects data on grizzly bear populations and publishes the Yellowstone Grizzly Bear Status Report. They also work to stop oil drilling and mining in the area, hold an annual scientific conference in Yellowstone National Park, and lobby Congress on environmental issues significant to the area.

PUBLICATION: *Greater Yellowstone Report* is a quarterly newsletter for members.

Greenpeace U.S.A.
1436 U Street, NW
Washington, DC 20009
(202) 462-1177
Founded: 1971

With more than 2 million members and an annual budget of over $50 million, Greenpeace is one of the largest, and most visible, environmental organizations in the world. Greenpeace was originally founded in Canada to oppose nuclear testing off Alaska's Amchitka Island and then expanded to protest nuclear testing across the globe. As the organization grew they began to diversify, and today they are involved in a wide range of environmental projects, including the protection of endangered land and sea animals and their habitats.

In some instances, Greenpeace advocates techniques with an ideology similar to that of Earth First!'s direct action policies. For example, Greenpeace ships have been known to block nuclear vessels from entering neutral ports and to ram commercial fishing vessels in order to destroy their drift net winches. Greenpeace members also are known to follow whaling vessels in small dinghies, placing themselves between whales and the harpoons. These types of activities have made the organization controversial on occasion.

PUBLICATION: A bimonthly magazine for members is called *Greenpeace*.

Hawk Mountain Sanctuary Association
Route 2
Kempton, PA 19529
(215) 756-6961
Founded: 1943

Fourteen species of birds migrate over Hawk Mountain in eastern Pennsylvania every year, including eagles, falcons, ospreys, and hawks. The Hawk Mountain Sanctuary was established in order to protect these birds and other wildlife. The association is dedicated to maintaining the sanctuary, organizing research projects, and operating a visitor's center and gift shop.

PUBLICATION: *The Hawk Mountain News* is an annual report sent to members.

The Izaak Walton League of America

1401 Wilson Boulevard, Level B
Arlington, VA 22209
(703) 528-1818
Founded: 1922

The Izaak Walton League is a domestic organization with 50,000 members and more than 400 local chapters across the country. The league was organized to promote clean air, soil, and water, and to protect wilderness and wildlife in the United States by educating the public, working with federal, state, and local governments, and using the courts to ensure that existing regulations are enforced. The league has also produced a weekly half-hour television show highlighting current environmental topics since 1975.

Some of the league's programs include Save Our Streams (SOS), which was organized in 1969 to monitor water quality across the United States on a local level. Campouts in California teach participants about the desert ecosystem. A National Conservation Center outside Washington, D.C., provides demonstrations on conservation techniques for visitors, and a youth conservation education program called Uncle Ike teaches children about conservation.

PUBLICATIONS: *Outdoor America* is a quarterly magazine for members. *Splash* is a quarterly newsletter for the Save Our Streams program.

Land Trust Alliance

900 Seventeenth Street NW, Suite 410
Washington, DC 20006
(202) 638-4725
Founded: 1982

Land Trust Alliance promotes public awareness and the creation of land trusts in which private lands are set aside by the owners to preserve segments of the wilderness.

PUBLICATIONS: *Starting a Land Trust* and *Conservation Easement Handbook* are informational pamphlets. The organization also published *National Directory of Conservation Land Trusts*.

League of Conservation Voters (LCV)
1150 Connecticut Avenue, NW, Suite 201
Washington, DC 20036
(202) 785-8683
Founded: 1970

The League of Conservation Voters (LCV) was founded by Marion Edey, who was working on the staff of a congressional committee at the time. The goal of the league is to influence voters to elect conservation-minded politicians. The league contributes funds to the campaigns of candidates with proenvironment records. The LCV has also produced and run television ads, and they publish the *National Environmental Scorecard,* which reports how members of Congress voted on environmental issues, and ranks them according to their environmental voting record.

PUBLICATIONS: The *National Environmental Scorecard,* published once a year, is available for a small fee. *Greengram* is a 2-page monthly newsletter on LCV activities and issues. The 12-page *Environmental Election Report* is published every other year shortly after the elections, listing LCF candidates and how they fared in the elections.

LightHawk
P.O. Box 8163
Santa Fe, NM 87504-8163
(505) 982-9656
Founded: 1974

LightHawk's primary tactic in its battle against environmental mismanagement is to use aircraft to fly over deforested and environmentally damaged areas, and report on their findings. They also take Congressmen and other public and community leaders on special fact-finding flights.

PUBLICATIONS: *Lighthawk: The Wings of Conservation* is a quarterly newsletter for members.

National Audubon Society
950 Third Avenue
New York, NY 10022
(212) 979-3000
Founded: 1905

The National Audubon Society was originally formed when 35 separate groups joined forces to oppose the killing of birds for their decorative feathers. The society is still known for their efforts to protect birds. Over the past 90 years, however, the society has evolved to represent a multitude of environmental concerns, focusing on the conservation of plants, animals, and habitats. The society purchases lands and operates sanctuaries and nature centers, lobbies Congress, offers educational programs and ecology camps and workshops, and conducts conservation research. There are more than 600,000 members of the society today, with over 500 local chapters.

PUBLICATIONS: *Audubon* is a bimonthly magazine for members. *Audubon Activist* is a bimonthly news journal for members of the society's Audubon Activist Network.

National Parks and Conservation Association
1776 Massachusetts Avenue, NW
Washington, DC 20007
(202) 223-6722
Founded: 1919

This association was created to promote the U.S. National Park System. They have worked successfully to expand the boundaries of many existing parks in the United States by lobbying Congress and by purchasing lands adjacent to national parks and then donating that land to the park system. The association also strives to educate the public, and initiated a program to educate elementary school students on endangered species and habitats. Under the park-watcher program, volunteers alert the association to any threat to a national park.

PUBLICATIONS: *National Parks* is a bimonthly magazine for members. *Exchange* is a bimonthly newsletter for volunteers in the park-watcher program.

National Wildlife Federation (NWF)
1400 Sixteenth Street, NW
Washington, DC 20036-2266
(202) 797-6800
Founded: 1936

With 5.6 million members, the National Wildlife Federation (NWF) is the largest environmental-education organization in the United States. The federation was originally founded as the General Wildlife Federation by President Franklin D. Roosevelt at the first North American Wildlife Conference in 1936. The name was revised to the National Wildlife Federation in 1938. The original and continuing goal of the federation is to protect and conserve the "vanishing wildlife resources of a continent" through education, research, cooperation with government agencies and private organizations, and legal action.

NWF projects include educational Wildlife Camps for children, and Conservation Summits for children and adults. The federation sponsors

National Wildlife Week once a year, sending information kits and posters to the nation's classrooms to promote awareness.

On the legal side of the organization, NWF won a lawsuit to force the EPA and the Department of the Interior to implement natural resource damage provisions of the Superfund. Another lawsuit halted coal leasing in the Fort Union area of North Dakota until the federal coal program was reformed.

The NWF operates seven resource conservation centers across the country, as well as the Laurel Ridge Conservation Education Center in Vienna, Virginia. The Laurel Ridge facility includes a nature trail accessible to the physically challenged, and a conservation library with over 10,000 titles.

PUBLICATIONS: Members receive either *National Wildlife* or *International Wildlife,* or both for an extra fee. *Ranger Rick* magazine is for children aged 6 to 12. *Your Big Backyard* is for preschoolers.

Native Forest Council
P.O. Box 2171
Eugene, OR 97402
(503) 688-2600
Founded: 1988

Outrage over what many considered excessive logging in the national forests during the late 1980s led to the creation of this group, which works to protect the remaining forests in the United States through education, legislation, and lobbying.

PUBLICATIONS: *Forest Voice* is a quarterly newspaper, distributed to more than 1 million people since 1988.

Natural Resources Defense Council (NRDC)
40 West Twentieth Street
New York, NY 10011
(212) 727-2700
Founded: 1970

The Natural Resources Defense Council (NRDC) was formed by a group of Yale Law School classmates to fight for environmental causes through lobbying and litigation. Today the organization has 168,000 members and five regional offices, including New York, Honolulu, Los Angeles, San Francisco, and Washington, D.C. The NRDC has won lawsuits protecting lakes and streams, fighting coal and oil leases, and enforcing the Clean Air Act. Currently the organization is trying to protect Alaska's National Arctic Wildlife Refuge from oil development.

PUBLICATIONS: The NRDC published *The Rainforest Book* in 1990. *The Amicus Journal* is a quarterly magazine for members. The *NRDC Newsline* is a newsletter published five times per year with updates on the organization's operations.

The Nature Conservancy
1815 North Lynn Street
Arlington, VA 22209
(703) 841-5300
Founded: 1951

The Nature Conservancy is an organization with a somewhat unique and highly successful approach to conservation. While many organizations purchase lands to set aside and preserve, The Nature Conservancy makes acquiring land its number one goal. The conservancy today has over 700,000 members and an annual budget of over $156 million. It is the richest environmental organization in the country and currently owns and manages 1.3 million acres of land, valued at over $400 million. Included in these holdings are 1,300 preserves from Santa Cruz Island off the coast of California to Tallgrass Prairie Preserve in Oklahoma, and forests and wetlands throughout New England, New Mexico, California, Nevada, and Oregon. In 1989 the conservancy purchased an additional 40,000 acres for Adirondack Park in New York, and in 1990 they completed the largest private land acquisition in conservation history when they spent $18 million on the purchase of the 321,000 acre (502 square miles) Gray Ranch in southwestern New Mexico.

The Nature Conservancy works closely with government agencies, including the BLM, the USFWS, and the USFS. One of the conservancy's techniques includes purchasing land to swap with these agencies for federal land that the conservancy feels should be protected. Also, in some cases the U.S. agencies express interest in saving certain lands, but are unable to move fast enough due to their extensive bureaucracies. In such cases The Nature Conservancy will purchase the lands, hold them until the government agency clears enough funds, and then sell the land to the government agency. Since 1953 they have protected 7.5 million acres in the United States through such techniques, and have helped to protect 20 million acres outside the United States.

PUBLICATIONS: *Nature Conservancy* is a quarterly magazine for members.

North American Wildlife Foundation (NAWF)
102 Wilmot Road, Suite 410
Deerfield, IL 60015
(708) 940-7776
Founded: 1911

Wildlife preservation, wetlands management, and soil and water conservation are the first priorities of the North American Wildlife Foundation

(NAWF). The foundation established and maintains the Delta Waterfowl and Wetlands Research Station on the Delta Marsh in Manitoba, Canada. Graduate students are able to study and conduct research projects at the station relating to wetlands and waterfowl with an aim toward protecting these resources. The foundation is also active in agricultural research, concentrating on land use of North American Prairies.

PUBLICATIONS: Members receive the *Annual State of Prairie Ducks Report,* and the quarterly *Waterfowl Report.*

Public Forestry Foundation
P.O. Box 371
Eugene, OR 97440-0371
(503) 687-1993

Conservation of forest resources is the number one concern of the Public Forestry Foundation. The group works toward managing timber harvesting to sustainable levels. They monitor timber sales and conduct seminars for the public and professionals in the field of forestry.

PUBLICATIONS: The *Citizen Forester* is a quarterly newsletter.

Save America's Forests
4 Library Ct. SE
Washington, DC 20003
(202) 544-9219

With an office almost directly across the street from the U.S. Capitol, Save America's Forests works to introduce and support legislation aimed at more ecologically oriented management of the nation's forests, such as the proposed Forest Biodiversity and Clearcutting Prohibition Act, HR 1164. Save America's Forests is actually a coalition of 110 conservation groups and 500,000 individual members that share a concern for forests and the plant and animal species within them.

PUBLICATIONS: *Save America's Forests* is a periodic newsletter.

Save-The-Redwoods League
114 Sansome Street, Room 605
San Francisco, CA 94104
(415) 362-2352
Founded: 1918

The Save-The-Redwoods League was founded to purchase redwood forests in order to preserve them. Once the League makes a purchase it donates the lands to one of 35 state and federal parks in California. Since its founding, the league has purchased and donated more than 260,000 acres. The league also uses its clout to lobby the USFS for more protection for redwoods and giant sequoia groves.

PUBLICATIONS: Two bulletins per year are sent to members. Educational pamphlets are also published occasionally.

Sierra Club
P.O. Box 7959
San Francisco, CA 94120-9943
(415) 776-2211
Founded: 1892

The Sierra Club is among the best known and most influential environmental organizations in the world. Naturalist John Muir founded the Sierra Club over a century ago to protect his beloved Sierra Nevada Mountains. Since that time, the organization's goal has expanded to protecting the natural environment of the entire United States and the world, with 650,000 members and 57 local chapters nationwide.

The Sierra Club's methods include environmental research, education of the public through publications, nature trips, outings, meetings, and political lobbying. The club's areas of interest include wildlife and wilderness preservation, national parkland acquisitions, water and energy resources, population, toxic waste, and pollution. The club also encourages the enjoyment of the natural world and leads approximately 250 or more trips each year to destinations around the world.

PUBLICATIONS: *Sierra* is a quarterly magazine for all members. Local chapters also publish their own newsletters. The Sierra Club also publishes several calendars and an extensive list of books each year.

Sierra Club Legal Defense Fund
2044 Fillmore Street
San Francisco, CA 94115
(415) 567-6100
Founded: 1971

The Sierra Club Legal Defense Fund is an organization independent from the Sierra Club, but they do work with the Sierra Club as well as almost every other major environmental group to provide legal representation for environmental causes. The fund has offices in San Francisco, Washington, D.C., Denver, Juneau, Seattle, and Honolulu.

The Defense Fund successfully blocked the development of a ski resort in California's Mineral King Valley in what is now part of Sequoia National Park. They also stopped construction of the world's largest coal-fired power plant in Kaiparowits Plateau, Utah, defeated oil and gas drilling projects proposed by James Watt in Wyoming, and work to prevent clear cutting in sensitive areas of the Pacific Northwest.

PUBLICATIONS: A quarterly newsletter, *In Brief*, is available to supporters.

Waterfowl U.S.A., Limited
Box 50, The Waterfowl Building
Edgefield, SC 29824
(803) 637-5767

Waterfowl U.S.A. is an organization of hunters and outdoor sports enthusiasts with the common goal of preserving America's waterfowl habitats. The group donates money for wetland acquisition and protection and sponsors conservation projects, including the creation of waterfowl refuges.

PUBLICATIONS: *Waterfowl* is a bimonthly magazine. Along with the subscription, members also receive a waterfowl U.S.A. watch and duck call.

Whitetails Unlimited (WTU)
P.O. Box 422
Sturgeon Bay, WI 54235
(414) 743-6777
Founded: 1982

Whitetails Unlimited (WTU) was founded by deer hunters to promote sound deer management in the United States, among both hunters and nonhunters alike, through research and education. The organization has started over 100 local chapters across the country. Education programs include WTU-sponsored lectures and seminars, brochures, pamphlets, and other published materials. WTU financially supports research on deer by federal and state governments and private universities.

PUBLICATIONS: Members receive *The Deer Trail* magazine, published quarterly, and have access to the Whitetails Unlimited video library. *Trail Talk* is a tabloid published once a year.

The Wilderness Society
900 Seventeenth Street, NW
Washington, DC 20006-2596
(202) 833-2300
Founded: 1935

The goal of the Wilderness society is the protection and preservation of America's forests, parks, rivers, deserts, and shore lands. The society is one of the largest and most influential environmental organizations in the nation, with over 300,000 members and 15 field offices throughout the United States. It is the only national conservation devoted primarily to public lands protection and management issues.

In order to achieve its goals, The Wilderness Society lobbies congress extensively for legislation to protect the environment. The society has been involved in almost all major legislation relating to the public lands for the past 50 years, including the Wilderness Act (1964), Endangered Species Act (1973), Alaska Lands Act (1980), and Tongass Timber Reform Act (1990).

PUBLICATIONS: *Wilderness* is a quarterly magazine for members.

Wildlife Conservation Society
New York Zoological Society
185th Street and Southern Boulevard
Bronx, NY 10460-1099
(718) 220-5895
Founded: 1895

Affiliated with the New York Zoological Society, Wildlife Conservation International is among the oldest conservation organizations in the country. Their aim is to preserve endangered ecosystems and species. To achieve this, the organization conducts scientific studies and research projects.

PUBLICATIONS: The organization publishes *Wildlife Conservation Magazine* bimonthly.

The Wildlife Society
5410 Grosvenor Lane
Bethesda, MD 20814
(301) 897-9770
Founded: 1936

This organization was originally called The Society of Wildlife Specialists, but changed the name to The Wildlife Society in 1937. The majority of the 8,500 members are wildlife management professionals from around the world. The goals of the society are to promote sound management of wildlife and the environment and to educate the public on wildlife issues. The organization achieves these goals primarily through its publications and through programs with schools, such as their Student Wildlife Conclaves, which are held across the country for university students. The conclaves feature seminars, guest speakers, field trips, and wildlife quiz bowls.

On the professional level, the Wildlife Society has a certification program in which a wildlife professional can become an Associate Wildlife Biologist, with the proper education, or a Certified Biologist, with education and experience. The society also gives annual awards, such as the Aldo Leopold Memorial Award, for outstanding accomplishments in the field.

PUBLICATIONS: The *Journal of Wildlife Management,* published quarterly, and the monthly *Wildlife Society Bulletin* are scientific journals for wildlife professionals, professors, and students. *The Wildlifer* is a bimonthly newsletter for members.

World Wildlife Fund (WWF)
1250 Twenty-fourth Street, NW
Washington, DC 20037
(202) 293-4800
Founded: 1961

With over 1 million members, the World Wildlife Fund (WWF) is one of the largest environmental organizations in the United States. In addition, WWF has a network of affiliated organizations in 23 other countries. This makes WWF the largest international environmental organization in the world.

The WWF has nine official goals which it strives to fulfill: protect habitat, promote ecologically sound development, support scientific investigation, promote education in developing countries, provide training for local wildlife professionals, encourage self-sufficiency in developing countries, monitor international wildlife trade, and influence public opinion and the policies of governments and private institutions.

The WWF has worked with foreign governments to help create national parks in places such as Nepal, Kenya, and Peru. All told, the WWF has initiated over 1,600 projects throughout the world, including a program known as TRAFFIC, which monitors both legal and illegal international trade of wildlife and plants, and is aimed at stopping the illegal trade.

PUBLICATIONS: *Focus* is a bimonthly newsletter for members.

The Worldwatch Institute
1776 Massachusetts Ave., NW
Washington, DC 20036
(202) 452-1999
Founded: 1975

The Worldwatch Institute was founded to educate the general public and the policy makers in Washington, D.C., of the relationship between environmental issues and the economy. The institute studies both global and domestic issues and publishes *State of the World,* an annual guide for government officials, environmentalists, students, and professionals. The book is printed in 12 languages and sells around 200,000 copies per year. The institute has also produced a ten-part series on "Nova" with WGBH-TV Boston based on the State of the World.

PUBLICATIONS: A paperback copy of *State of the World* is included with membership. Additional reports and papers are also published on occasion.

7

Selected Print Resources

THIS CHAPTER CONTAINS AN annotated bibliography listing books with information on wilderness issues in America. The listings are divided into eight catagories, including General, Biographies, America's Forests, America's Rivers and Lakes, America's Prairies, America's Deserts, America's Wetlands, and Alaskan Wilderness. The books included here were chosen to provide the best range of information on each topic for students from junior high school to college, and beyond.

General

Amdur, Richard. **Wilderness Preservation.** New York, NY: Chelsea House, 1993. 110 p. ISBN 0-7910-1580-7; 0-7910-1605-6 (paperback).

Written for junior high school and high school age readers, this book provides a good introduction to wilderness preservation, including where wilderness is located and how it is managed in the United States.

America's Hidden Wilderness. Washington, DC: National Geographic, 1988. 199 p. ISBN 0-87044-666-5 (regular edition); 0-87044-671-1 (library edition).

Much of the wilderness in North America is not in national parks or official wilderness areas. This book, written and published by the National Geographic Society, explores seven of these spots, including five in the United States and one each in Canada and Mexico. The U.S.

locations consist of the Mojave Desert in California; Baxter State Park, Maine; Utah's Grand Gulch; Arctic National Wildlife Refuge in Alaska; and Great Burn on the Idaho-Montana Border. Maps and extensive color photographs accompany the text.

Chandler, William J., ed. **Audubon Wildlife Report.** San Diego: Academic Press, 1993. 585 p.

This biannual report includes highly informative articles on current topics concerning the preservation of wildlife in the United States. Each report focuses on the issues surrounding a specific federal agency. Past volumes have featured the Army Corps of Engineers, USFWS, USFS, BLM, and National Marine Fisheries Service.

Chase, Alston. **Playing God in Yellowstone: The Destruction of America's First National Park.** New York: Harcourt Brace Jovanovich, 1987. 464 p. ISBN 0-15-672036-1.

This book documents how mismanagement in Yellowstone National Park has led to imbalances in the natural species. For example, when managers carried out a plan to eradicate wolves from the park, this led to an explosion in the elk population.

Curry-Lindahl, Kai. **Conservation for Survival—An Ecological Strategy.** New York: William Morrow, 1972. 335 p. No ISBN.

This is an informative survey of ecological problems around the world. Separate sections discuss all of the major problems with air, water, soil, vegetation, animals, and man. Another section looks at the most pressing problems of each specific continent. The final chapters look to the future and what can be done to manage our planet for future generations.

Donlan, Edward F. **The American Wilderness and Its Future: Conservation Versus Use.** New York: Franklin Watts, 1992. 144 p. ISBN 0-531-11062-1.

This book is a thorough introduction to the controversies surrounding wilderness conservation in the United States, written for a young adult audience. Conflicts between economic and environmental interests in such areas as logging, pollution, and development are discussed and explained. The National Park and National Forest systems in the United States are also covered thoroughly. This book is an excellent source for an initial look into the complex issues of American conservation.

Dubasak, Marilyn. **Wilderness Preservation: A Cross-Cultural Comparison of Canada and the United States.** New York: Garland, 1990. 237 p. ISBN 0-8240-2517-2.

The history of conservation and preservation in Canada and the United States is presented, with a look at the differences in management, organization, and thought.

Ehrlich, Paul, and Anne Ehrlich. **Extinction—The Causes and Consequences of the Disappearance of Species.** New York: Random House, 1981. 305 p. ISBN 0-394-51312-6.

Extinction, from the dinosaurs to modern day species is discussed, with an emphasis on animals driven to extinction by mankind in the twentieth century. The book, one of the best available on the subject, shows how man drives species to extinction, what the consequences are, and what we can do to save the remaining plants and animals.

Foreman, Dave. **Confessions of an Eco-Warrior.** New York: Harmony, 1991. 229 p. ISBN 0-517-58123-X.

Part autobiography, part historical overview of conservation, this book touches on a variety of subjects from the author's point of view. Several chapters are devoted to the practice of monkeywrenching. Above all, the purpose of the book is to encourage people to think about how they can get involved in saving the wilderness, or, as the author states "to get you, the reader, thinking and wondering, whether you like it or not. What is important is that you do something. Now."

Foreman, Dave, and Howie Wolke. **The Big Outside: A Descriptive Inventory of the Big Wilderness Areas of the United States.** New York: Harmony, 1992. 499 p. ISBN 0-517-58737-8.

Foreman and Wolke provide maps depicting the large roadless areas in each state, describe them in detail, and then discuss the status of the areas in regard to external threats. The text includes a history of roadless area protection and designation and an appendix list of the roadless areas.

Freemuth, John C. **Islands under Siege: National Parks and the Politics of External Threats.** Lawrence: University Press of Kansas, 1991. 186 p. ISBN 0-7006-0434-0.

Threats both in and around the national parks are addressed, including smog, overcrowding, water pollution, mineral extraction, and development. Freemuth also discusses possible solutions to these problems.

Frome, Michael. **National Park Guide.** New York: Rand McNally, 1993. 247 p. No ISBN.

This highly informative vacation guide is printed annually, and contains several pages on each national park in the United States. The guide is designed for people who are planning a vacation, and includes tourist information, but it also provides a history of the establishment of each park and in many cases explains the geology of the region and the types of wildlife that can be found there.

————. **Regreening the National Parks.** Tucson: University of Arizona Press, 1992. 289 p. ISBN 0-8165-0956-5; 0-8165-1288-4 (paperback).

This travel writer is well-known for his annual guide to the national parks (above). Here, in *Regreening the National Parks,* he takes a different look at the parks, examining the politics of their management, from the role of the director on down to the ranger in the field. Frome gives an inside view on how struggles between preservationists and industrialists, concessionaires, and ranchers often leave the parks worse for the wear, and he offers thoughts on correcting some of these problems.

Hartzog, George B. **Battling for the National Parks.** Mt. Kisco, NY: Moyer Bell, 1988. 284 p. ISBN 0-918825-70-9.

National parks that had a tough fight making it through the bureaucratic process are covered with a look at the battles they endured in Congress. These include Redwood National Park, The Everglades, several Alaskan parks, and others. The book concentrates on the time period between the early 1960s and 1980s.

Hendee, John C., George H. Stankey, and Robert C. Lucus. **Wilderness Management.** Golden, CO: North American Press, 1990. 546 p. ISBN 1-55591-900-6.

This is among the most comprehensive texts written to date on wilderness management in the United States. Topics include history, philosophy, management concepts, laws, wildlife, wildfires, recreation, and the use of wilderness areas.

Irland, Lloyd C. **Wilderness Economics and Policy.** Lexington, MA: Lexington, 1979. 225 p. ISBN 0-6690-2821-5.

United States preservation policy is discussed from an economic point of view. The author is a professor of forest economics and presents his argument evenly, not weighing in favor or against preservation.

Lawrence, Bill. **The Early American Wilderness: As the Explorers Saw It.** New York: Paragon House, 1991. 245 p. ISBN 1-5577-8145-1.

Lawrence begins in about A.D. 800 with the Vikings and continues until the Lewis and Clark expedition in 1804. He provides a history of all of the great explorations into the American wilderness in between, including the expeditions of Champlain, De Soto, La Salle, Coronado, Radisson, and others.

Lowe, David W., John R. Mathews, and Charles J. Moseley, eds. **The Official World Wildlife Fund Guide to Endangered Species of North America.** Washington, DC: Walton Beacham, 1990. 1180 p. ISBN 0-933833-17-2.

This excellent two-volume encyclopedia is updated every two years and includes a two-page listing for every endangered species in North America. The information includes photographs, maps, and detailed text describing physical characteristics, behavior, habitat, range, and an outlook for recovery.

McHenry, Robert, with Charles Van Doren, eds. **A Documentary History of Conservation in America.** New York: Preaeger, 1972. 422 p. No ISBN.

Included in this work are poems and short clips from novels, speeches, and other writings that relate to conservation. There is a wide range of genre and history included, ranging from Plato (ca. 380 B.C.) and Aeschylus (ca. 460 B.C.), to Percy Bysshe Shelly, and other writings that portray thoughts and opinions on environmental topics from different eras in U.S. history.

Matthiessen, Peter. **Wildlife in America.** New York: Viking, 1987. 332 p. ISBN 0-6708-1906-9.

The author describes wildlife across America, in this updated version of a now classic book, originally published in 1959. Color drawings and black and white photographs accompany the text. An appendix lists endangered species and legislation affecting wildlife in the United States.

Mitchell, John G. **Dispatches from the Deep Woods.** Lincoln: University of Nebraska Press, 1991. 304 p. ISBN 0-803-23146-6.

The author describes the different types of forests in the United States, the environmental problems they face, and possible solutions to those problems.

Naar, John, and Alex J. Naar. **This Land Is Your Land: A Guide to America's Endangered Ecosystems.** New York: HarperCollins, 1993. 388 p. ISBN 0-06-05165-8; 0-06-096387-5 (paperback).

Every type of ecosystem in North America is covered in this excellent overview of environmental problems, from rivers, lakes, and wetlands, to forests, deserts, and prairies. The book goes into detail on controversies and environmental challenges in specific areas, and offers suggestions for what the reader can do to help.

Nash, Roderick. **Wilderness and the American Mind.** New Haven, CT: Yale University Press, 1982. 425 p. ISBN 0-300-02905-5; 0-300-02910-1 (paperback).

This book is a classic among conservation literature, originally published in 1967, and is required reading in many university courses. It presents a history of conservation in the United States and points to a gradual change in the way Americans think about the wilderness and what it means to them.

————, ed. **The American Environment: Readings in the History of Conservation.** Reading, MA: Addison-Wesley, 1968. 236 p. No ISBN.

This anthology is a collection of influential essays, speeches, and other writings on the environment by influential Americans, from "An Artist Proposes a National Park," by George Catlin (1832), to "Beautification," by Lyndon B. Johnson (1964). Included are writings by such men as Frederick Law Olmsted, Theodore Roosevelt, Gifford Pinchot, Aldo Leopold, and Robert Marshall.

The New America's Wonderlands: Our National Parks. Washington, DC: National Geographic, 1980. 463 p. ISBN 0-87044-332-4.

This is a comprehensive look at the national parks dedicated before 1980, as well as selected national monuments, seashores, lakeshores, recreation areas, and scenic rivers. The book is heavily illustrated with exceptional color photographs.

Norton, Bryan G., ed. **The Preservation of Species: The Value of Biological Diversity.** Princeton, NJ: Princeton University Press, 1986. 305 p. ISBN 0-6910-8389-4.

This series of 11 separate articles discusses human-caused extinction, the reasons behind it, and what can be done to preserve endangered species. The articles touch on economics, property rights, biology, and philosophy.

Oelschlaeger, Max. **The Idea of Wilderness: From Prehistory to the Age of Ecology.** New Haven, CT: Yale University Press, 1991. 477 p. ISBN 0-300-0851-3.

Environmental philosophy is examined historically, with a look at thoughts and ideas from paleolithic cultures through more recent men such as John Muir, Henry David Thoreau, and Aldo Leopold.

Repetto, Robert, and Malcolm Gillis, eds. **Public Policies and the Misuse of Forest Resources.** New York: Cambridge University Press, 1990. 432 p. ISBN 0-521-34022-5; 0-521-33574-4 (paperback).

The management policies of the USFS and other federal agencies are held accountable for unnecessary destruction of timber in the national forests.

Smith, Darren L. **Parks Directory of the United States: A Guide to 3,700 National and State Parks, Recreation Areas, Historic Sites, Battlefields, Monuments, Forests, Preserves, Memorials, Seashores, and Other Designated Recreation Areas in the United States Administered by National and State Park Agencies.** Detroit: Omnigraphics, 1992. 525 p. ISBN 1-558-88765-2.

Each park listed in this guide is described with information on the size, location, camping and recreational activities, fees, hours of operation, etc.

Smith, Frank. **The Politics of Conservation—The First Political History of the Conservation and Development of America's Natural Resources.** New York: Pantheon Books, 1966. 338 p. No ISBN.

This book covers major legal battles in the history of conservation in America. It starts with the very first session of Congress, in 1789, and works its way through congressional and presidential legislation up until the 1960s.

Tilden, Freeman. **The National Parks: The Classic Book on the National Parks, National Monuments, and Historic Sites.** 3d. rev. ed. New York: Knopf, 1986. 603 p. ISBN 0-3947-4294-X.

Originally published in 1968, this history and guidebook was one of the first to contain comprehensive information on the national parks. This version has been updated and revised by Paul Schullery.

Udall, Stewart L. **The Quiet Crisis and the Next Generation.** Layton, UT: Gibbs Smith, 1988. 298 p. ISBN 0-87905-333-X.

Stewart Udall served as Secretary of the Interior under the Kennedy and Johnson administrations. He originally published *The Quiet Crisis* in 1963. This updated version is a history of wilderness conservation and preservation in the United States, with a look at contributions of important individuals.

Weinstock, Edward B. **The Wilderness War: The Struggle to Preserve Our Wildlands.** New York: Julian Messner, 1982. 191 p. ISBN 0-671-42246-4.

Weinstock provides a history of wilderness preservation and destruction in the United States, from the beginning of westward expansion through the 1970s. He discusses early preservationists, the USFS, the Wilderness Act, and the effects of modern industry.

Wilderness America: A Vision for the Future of the Nation's Wildlands. Washington, DC: The Wilderness Society, 1989. 64 p.

This brief, yet beautifully produced book, is full of maps and spectacular color photographs, and examines different wilderness regions throughout the United States. Recommendations are made on how to best manage them for the future.

Zaslowsky, Dyan, and The Wilderness Society. **These American Lands.** New York: Henry Holt, 1986. 404 p. ISBN 0-8050-0084-4.

The history of public lands in the United States is documented and explained in this informative and well-organized text. Covered in separate chapters are the National Park System, National Forest System, BLM, National Wildlife Refuge System, National Wilderness Preservation System, National Interest Lands of Alaska, National Trails Systems, and the National Wild and Scenic Rivers.

Biographies

Brower, David. **For Earth's Sake: The Life and Times of David Brower.** Salt Lake City: Peregrine Smith, 1990. 556 p. ISBN 0-87905-013-6.

This is a very well executed biography of David Brower, one of the most influential men in the conservation movement.

Clepper, Henry. **Leaders of American Conservation.** New York: Ronald Press Company, 1971. 353 p. No ISBN.

Short biographies (150–250 words) of American figures important to the conservation movement in the United States. The majority of those listed are men and women who have devoted their lives to wilderness conservation by running environmental organizations, publishing journals, and working for private and government conservation projects.

Strong, Douglas H. **Dreamers and Defenders: American Conservationists.** Lincoln: University of Nebraska Press, 1988. 295 p. ISBN 0-8032-4161-5; 0-8032-9156-6 (paperback).

A history of conservation in the United States with chapters focusing on the contributions of ten specific individuals, including Gifford Pinchot, John Muir, Aldo Leopold, and Franklin Delano Roosevelt. This book was originally published in 1971 under the title *The Conservationists*.

Vickery, Jim Dale. **Wilderness Visionaries.** Merrillville, IN; ICS Books, 1986. 263 p. ISBN 0-934802-27-0.

This book limits its scope to six biographies: Henry David Thoreau, John Muir, Robert W. Service, Robert Marshall, Calvin Rustrum, and Sigurd F. Olson. The author states that he chose these six "because their greatness is beyond question" and "their spirit, moreover, remains with us to this day."

Wild, Peter. **Pioneer Conservationists of Eastern America.** Missoula, MT: Mountain Press, 1986. 280 p. ISBN 0-87842-126-2; 0-87842-149-1 (paperback).

This book begins with a look at the development of the conservation movement in the United States and then moves on to 10- to 15-page biographies of 15 influential conservationists in the history of the East Coast, from the early days of George Marsh and George Grinnell to Ralph Nader and the present day.

————. **Pioneer Conservationists of Western America.** Missoula, MT: Mountain Press, 1979. 246 p. ISBN 0-87842-107-6.

Similar in style to *Pioneer Conservationists of Eastern America,* this book covers such westerners as John Muir, Gifford Pinchot, Stephen Mather, Enos Mills, David Brower, and Stewart Udall.

America's Forests

Beum, Frank R. **A Time for Commitment: Case-Study Reviews of National Forest Wilderness Management.** Washington, DC: The Wilderness Society, 1990. 50 p. No ISBN.

Beum studies four wilderness areas: Ansel Adams, California; Superstition, Arizona; Maroon Bells–Snowmass, Colorado; and Great Gulf, New Hampshire to determine whether or not they have been effectively managed since the passing of the Wilderness Act in 1964. His findings indicate that not enough is being done to halt degradation of these lands.

Clary, David. **Timber and the Forest Service.** Lawrence: University Press of Kansas, 1986. 252 p. ISBN 0-7006-0314-X.

This is a historical examination of timber management in the United States and the development of the USFS. The book provides a great deal of information on the subject of America's national forests.

Cubbage, Frederick W., Jay O'Laughlin, and Charles S. Bullock, III. **Forest Resource Policy.** New York: Wiley, 1993. 562 p. ISBN 0-471-62245-1.

This college-level text book is an excellent study of federal forest management policies. It covers the extent, location, and policy history of forests in the United States, both private and federal.

Dietrich, William. **The Final Forest: The Battle for the Last Great Trees of the Pacific Northwest.** New York: Simon and Schuster, 1992. 303 p. ISBN 0-671-72967-5.

The controversies of logging in the old-growth forests of the Pacific Northwest are described, including the ecology, economics, and politics. Deitrich devotes separate chapters to interviews of a biologist, trucker, logging representative, USFS representative, and others.

Ervin, Keith. **Fragile Majesty: The Battle for North America's Last Great Forest.** Seattle: Mountaineers, 1989. 272 p. ISBN 0-89886-230-2; 0-89886-204-3 (paperback).

Ervin examines the ecology of the Pacific Northwest old-growth forests, and provides an in-depth look at the effects of logging in the region from an environmental point of view.

Fuller, Margaret. **Forest Fires: An Introduction to Wildland Fire, Behavior, Management, Firefighting and Prevention.** New York: Wiley, 1991. 238 p. ISBN 0-4715-2189-2.

This text covers fires during the 1980s in North America, Australia, and China, including the Yellowstone fires. The book explains fire types and behavior, the role of weather, types of fuels, aftereffects, fire prevention, management, and fire fighting. This is one of the most comprehensive books to date on the topic of forest fires and fire management.

Holland, I. I., G. L. Rolfe, and David Anderson. **Forests and Forestry.** Danville, IL: Interstate, 1990. 476 p. ISBN 0-813-42854-2.

A thorough introduction to the topic of forestry, this textbook discusses forest ecology and biology, forest products and economics, artificial reforestation, resource management, wood types and uses, harvesting, and forest fires.

Johnson, Edward A. **Fire and Vegetation Dynamics: Studies from the North American Boreal Forest.** Cambridge, NY: Cambridge University Press, 1992. 129 p. ISBN 0-5213-4151-5.

From a scientific standpoint, this book explains the causes and mechanics of fire in the forest, including climate, fire intensity, fire behavior, and the effects of fire on the distribution and population of boreal trees.

Perlin, John. **A Forest Journey—The Role of Wood in the Development of Civilization.** New York: Norton, 1989. 445 p. ISBN 0-393-02667-1.

This is a history of mankind's use and abuse of forests throughout the world and the resulting problems. Those problems addressed include depletion of firewood supplies, flooding, loss of soil, desertification, and declining soil productivity, beginning with the use of wood during the bronze age and continuing through nineteenth-century America.

Roth, Dennis M. **The Wilderness Movement and the National Forests.** College Station, TX: Intaglia Press, 1988. 92 p. ISBN 0-944091-01-6.

This is a history of major wilderness legislation in the United States. It provides an overview of the changes in political thought over the decades on the subject of conservation and discusses the specific conservation laws passed by Congress and the circumstances that shaped them, including the impact of private organizations in the environmental movement, and the differing opinions of Senators and members of Congress since the 1950s.

Sholly, Dan R., with Steven M. Newman. **Guardians of Yellowstone: An Intimate Look at the Challenges of Protecting America's Foremost Wilderness Park.** New York: Morrow/Quill, 1993. 317 p. ISBN 0-688-09213-6; 0-688-12574-3 (paperback).

This is a look at the Yellowstone fires of 1988 through the eyes of the head ranger in charge of orchestrating firefighting operations. Ranger Dan Sholly tells the story not only of the fires and organizing fire fighting crews, but also of responding to a confused and angry public and the media. He also discusses and defends NPS fire management policies.

Simpson, Ross W. **The Fires of '88: Yellowstone Park and Montana in Flames.** Helena, MT: American Geographic/Montana Magazine, 1989. 80 p. ISBN 0-9383-1466-1.

The author covered the fires as a journalist and wrote this book before the last of the Yellowstone fires was even put out. The accounts are vivid and include both anecdotes and a serious look at the causes and effects of the fires, including a day-by-day chronology of all of the major fires, and how they started, spread, and eventually were brought under control.

Stoddard, Charles Hatch. **Essentials of Forestry Practice.** New York: Wiley, 1987. 407 p. ISBN 0-471-84237-0.

An overview of forestry management, composition and distribution of forests, silvicultural systems, logging, protecting forests from fire, insects, and disease, and organization and administration of forest programs.

Vogt, Gregory. **Forests on Fire: The Fight to Save Our Trees.** New York: Franklin Watts, 1990. 143 p. ISBN 0-5311-0940-2.

This is an excellent introduction to forest fires and the U.S. government's fire management policies, with special attention to the Yellowstone fires of 1988.

Williams, Michael. **Americans and Their Forests: A Historical Geography.** New York: Cambridge University Press, 1989. 599 p. ISBN 0-521-33247-8.

This text provides a history of white expansion throughout the United States and the resulting effect on the nation's forests, concentrating on the time period from 1600 to 1933. The author examines the uses and quantity of forest, and discusses arguments for and against preservation.

Young, Raymond A., and Ronald L. Geise, eds. **Introduction to Forest Science.** New York: Wiley, 1990. 586 p. ISBN 0-471-85604-5.

This college-level textbook covers all aspects of forest science around the world, including forest biology, ecology, wildlife, timber management, and products, with a chapter on career opportunities in the profession of forestry. Numerous charts, graphs, tables, and photographs are included.

America's Rivers and Lakes

Brown, Bruce. **Mountain in the Clouds: A Search for the Wild Salmon.** New York: Simon and Schuster, 1982. 239 p. ISBN 0-671-43583-3.

This is a well-written, thorough history of the disintegration of the habitat and population of the Pacific salmon, due to dams, logging operations, and overfishing. The book mixes hard facts with personal insight.

Buffone, Susan, and Carolyn Fulco. **Acid Rain Invades Our National Parks.** Washington, DC: National Parks and Conservation Association, 1987. 48 p. No ISBN.

This report describes acid rain and looks at the effects it has on the rivers, streams, lakes, and ecosystems in 12 specific national parks across the country, from Acadia National Park in Maine to Sequoia and Kings Canyon National Park in California.

Firor, John. **The Changing Atmosphere: A Global Challenge.** New Haven, CT: Yale University Press, 1990. 145 p. ISBN 0-3000-3381-8.

This book shows that the problems of acid rain are not limited to any one country, but are a global problem since the atmosphere is shared by all. Ozone depletion and global warming are also discussed.

Goldsmith, Edward, and Nicholas Hidlyard. **The Social and Environmental Effects of Large Dams.** San Francisco, CA: Sierra Club, 1984. 404 p. ISBN 0-87156-484-9.

This text analyzes the environmental effects of large-scale dams. It also discusses the politics of dams and provides suggestions for alternatives to large-scale dam building.

Hansen, Paul. **Acid Rain and Waterfowl: The Case for Concern in North America.** Arlington, VA: Izaak Walton League, 1987. 39 p. No ISBN.

This short, 39-page paperback book shows which areas are threatened by acid rain and how it effects waterfowl as they migrate across polluted habitats. Productivity and population, for example, are directly impacted by the acidified waters.

Howells, Gwyneth Parry. **Acid Rain and Acid Waters.** New York: Ellis Horwood, 1990. 215 p. ISBN 0-1300-4797-X.

This is part of a series published by Ellis Horwood on Environmental Management Science and Technology, with a look at the scope and area of the problem and possible solutions.

Kennedy, I. R. **Acid Soil and Acid Rain.** New York: Wiley, 1992. ISBN 0-4719-3404-6.

This book presents the research findings of studies on acidification of soil from acid rain, with an examination of nitrogen and sulphur levels and their effects on plant life.

Lucas, Eileen, Helen J. Challand, and Harriet Stubbs. **Acid Rain.** Chicago, IL: Children's Press, 1991. 128 p. ISBN 0-5160-5503-8.

Written for a younger audience (through high school), this introduction to the problems of acid rain is clear and easily understood.

Park, Chris. **Acid Rain: Rhetoric and Reality.** New York: Methuen, 1987. 272 p. ISBN 0-4169-2190-6; 0-4169-2200-7 (paperback).

This is a broad look at the topic, divided into four parts: The problem, science, technology (cures and remedies), and politics of acid rain in the United States and throughout the world.

Pringle, Lawrence P. **Rain of Troubles: The Science and Politics of Acid Rain.** New York: Macmillan, 1988. 121 p. ISBN 0-0277-5370-0.

This text covers all aspects of the acid rain issue, and the connection between scientific findings and political policy and legislation.

Regans, James L., and Robert W. Rycroft. **The Acid Rain Controversy.** Pittsburgh: University of Pittsburgh Press, 1988. 228 p. ISBN 0-8229-3582-1; 0-8229-5404-4 (paperback).

This is one in a Pittsburgh University Press series on policy and institutional studies. Both the environmental and political aspects of acid rain in the United States are addressed, with an examination of U.S. government policy concerning domestic and international sides of the issue.

Schmandt, Jurgen, Judith Clarkson, and Roderick Hilliar, eds. **Acid Rain and the Friendly Neighbors: The Policy Dispute between Canada and the United States.** Durham, NC: Duke University Press, 1988. 344 p. ISBN 0-8223-0870-3.

Originally published in 1985, this revised edition includes articles pertaining to the governmental policies of Canada and the United States and the dispute caused by the fact that emissions from U.S. cars and factories condense in clouds that traverse the border and rain in Canada.

Thomas, Bill. **American Rivers: A Natural History.** New York: Norton, 1978. 221 p. ISBN 0-393-08838-3.

Beautiful color photography accompanies this description of the rivers of America. River location and size, as well as information on local flora and fauna, are included for all regions of the continental United States and Alaska.

Tyson, Peter. **Acid Rain.** New York: Chelsea, 1992. 127 p. ISBN 0-7910-1577-7; 0-7910-1602-1 (paperback).

From a series titled Earth at Risk, this book provides an explanation of the problems of acid rain, its causes, effects, and possible solutions. The information is interesting, clear, and easy to understand. It is intended as an introduction to the issue for anyone from junior high school to adult. Numerous photographs add to the text.

America's Prairies

Brown, Lauren. **Grasslands.** New York: Knopf, 1985. 608 p. ISBN 0-394-73121-2.

This is one in an excellent series of Audubon Society Nature Guides, designed for use in identifying specific plant, animal, and insect species while in the field. Included are nearly 600 different species with color photographs, maps, and descriptive, informative text.

Collins, Scott L., and Linda L. Wallace, eds. **Fire in North American Tallgrass Prairies.** Norman: University of Oklahoma Press, 1990. 175 p. ISBN 0-8061-2281-1.

Separate scientific articles examine the effects of fire on tallgrass prairie ecosystems, including a historic look at the formation of America's grasslands 5 million years ago. This book is for the advanced student.

Curry-Lindahl, Kai. **Wildlife of the Prairies and Plains.** New York: Abrams, 1981. 232 p. ISBN 0-8109-1766-1.

Prairies and plains throughout the world are covered in this book, with one specific chapter on the ecology of plants and animals on the American plains. Stunning color photographs show these grasslands, along with bison, coyotes, deer, and many other animals that live there.

Madson, John. **Where the Sky Began.** Boston, MA: Houghton Mifflin, 1982. 321 p. ISBN 0-395-25718-2.

This is one of the best overviews of the North American prairies and their ecology before and after European settlement, with a description of the original native prairie that remains in preserves today.

Matthews, Anne. **Where the Buffalo Roam: The Storm Over the Revolutionary Plan to Restore America's Great Plains.** New York: Grove Weidenfeld, 1992. 193 p. ISBN 0-8021-1408-3.

When two college professors from New Jersey came up with an idea to set aside 139,000 square miles of plains as a preserve for bison, a wave of protest followed. In this book, the author examines this proposal, and looks at the state of the American plains today.

America's Deserts

Aitchinson, Stewart. **A Wilderness Called Grand Canyon.** Stillwater, MN: Voyageur, 1991. 127 p. ISBN 0-89658-149-7.

The author describes the geology and ecology of the Grand Canyon region, including flora and fauna, rocks and fossils, and threats to the ever shrinking wilderness. The text is accompanied by an impressive collection of photographs.

Awinger, Ann Haymond. **The Mysterious Lands: A Naturalist Explores the Four Great Deserts of the Southwest.** New York: Truman Talley, 1989. 388 p. ISBN 0-525-24546-4.

In the style of a personal journal, the author describes her travels through the American deserts, including informative discussions of the regions' plants and animals.

Dykinga, Hack W., and Charles Bowden. **The Sonoran Desert.** New York: Abrams, 1992. 167 p. ISBN 0-8109-3824-3.

This enchanting look at the Sonoran Desert contains perhaps the most stunning collection of photographs of the region published to date. Separate sections cover portions of the desert in California, Arizona, and Mexico. Accompanying text provides valuable insight into the desert ecology.

Louw, Gideon, and Mary Seely. **Ecology of Desert Organisms.** New York: Longman, 1982. 194 p. ISBN 0-582-44393-8.

Written for the advanced high school or college student, this book describes the ecology of desert ecosystems and the plants and animals that live there. It provides extensive scientific information about deserts both in the United States and throughout the world.

MacMahon, James A. **Deserts.** New York: Knopf, 1985. 640 p. ISBN 0-394-73139-5.

Another in the Audubon Society field guides, this book provides one of the best overviews of America's deserts and their plants, animals, and geology. Nearly 600 species of plants, animals, and insects are catalogued with color photographs, maps, and descriptive accompanying text.

Wagner, Frederic H. **Wildlife of the Deserts.** New York: Abrams, 1980. 231 p. ISBN 0-8109-1764-5.

Beautiful color photographs are spread throughout this fascinating look at different forms of life found in the world's deserts. The text explains how these insects and animals survive in the arid environment. This book is perfect for junior high school students through adults.

America's Wetlands

Dahl, T. E., and C. E. Johnson. **Wetlands: Status and Trend in the Conterminous United States, Mid–1970's to Mid–1980's.** Washington, DC: U.S. Fish and Wildlife Service, 1991. 28 p.

This text was written as a report to Congress. Although the report is not long, it contains the best information available on the extent of the disappearance of wetlands in the United States.

George, Jean Craighead. **Everglades.** New York: HarperCollins, 1995. ISBN 0-0602-12228-4.

The evolution of the Florida Everglades is described as well as the impact humans have had on the region and the problems facing the ecosystem in the future.

Kusler, Jon A., and Mary E. Kentula, eds. **Wetland Creation and Restoration.** Washington, DC: Island Press, 1990. 594 p. ISBN 1-55963-045-0.

This anthology is a collection of scientific reports on the status of wetlands throughout the United States, and the possibilities of man-made expansion and/or restoration of depleted areas.

Mitsch, William J., and James, G. Gosselink. **Wetlands.** New York: Van Nostrand Reinhold, 1993. 722 p. ISBN 0-4420-0805-8.

This scientific text is an excellent, detailed introduction to wetlands for the college-level student, including information on the politics, ecology, biology, and location of wetland areas.

Alaskan Wilderness

Brown, Dave, and Paula Crane. **Who Killed Alaska?** Far Hills, NJ: New Horizon Press, 1991. 336 p. ISBN 0-88282-069-9.

This book concentrates on environmental damage caused by oil companies in Alaska, following, from a personal view, the author's often fruitless battles against them.

Jacobson, Michael J., and Cynthia Wentworth, et al. **Kaktovik Subsistence: Land Use Values through Time in the Arctic National Wildlife Refuge Area.** Fairbanks, AK: U.S. Fish and Wildlife Service, Northern Alaska Ecological Services, 1982. 142 p.

Kaktovik is the only village located inside the Arctic National Wildlife Refuge. The village is inhabited by less than 200 natives, and is located on Barter Island, just off the Beaufort Seacoast. This book is a look at the people of Kaktovik and their way of life.

Miller, Debbie S. **Midnight Wilderness: Journeys in Alaska's Arctic National Wildlife Refuge.** San Francisco, CA: Sierra Club Books, 1990. 238 p. ISBN 0-0875-6715-6.

In 1975 the author, Debbie Miller, moved from California to Alaska and took a teaching job just south of the Arctic National Wildlife Refuge, in Arctic Village, population 125. Miller and her husband spent summers hiking through the refuge. This is a first-hand account of those journeys and a close up look at the region, with some discussions of the controversies and politics of the region.

Murray, John A., ed. **A Republic of Rivers: Three Centuries of Nature Writing from Alaska and the Yukon.** New York: Oxford University Press, 1990. 325 p. ISBN 0-19-506102-0.

This anthology is both interesting and informative. It includes excerpts written by explorers and naturalists visiting Alaska from 1741 to 1989. Topics include descriptions of early expeditions, native peoples, and Alaskan wildlife.

Oil Production in the Arctic National Wildlife Refuge: The Technology and the Alaskan Oil Context. Washington, DC: United States Congress, Office of Technology Assessment, 1989. 123 p. For sale through the U.S. Government Printing Office.

Much of this report deals with the technological aspects of drilling in the arctic environment, from building structures on ice and permafrost, to drilling operations, and oil transportation. Existing arctic oil production in the Prudhoe Bay area is examined as a comparison to possible operations in the ANWR. Estimates of the possible amount of oil in the region are also discussed, as well as potential environmental impacts.

Siy, Alexandra. **Arctic National Wildlife Refuge.** New York: Dillon Press, 1991. 80 p. ISBN 0-8751-8468-5.

Though written for elementary-aged children, this book does include both interesting and valuable information as an introduction to the ANWR. It also contains many excellent color photographs of the region and its wildlife.

Watkins, Tom H. **Vanishing Arctic: Alaska's National Wildlife Refuge.** Photographs by Wilbur Mills and Art Wolfe. New York: Aperture in association with The Wilderness Society, 1988. 86 p. ISBN 0-8938-1329-X.

This coffee table type book is filled with absolutely stunning photographs of the ANWR. The accompanying text is the author's first-person narration of a trip through the refuge. The oil drilling controversy is also discussed at the end of the book.

8

Selected Nonprint Resources

Following is a list of films and videos on the subject of wilderness preservation, with information on the content and source. The USFS has many films and videos available to borrow for only the cost of return postage. The NPS also has many excellent titles available at reasonable prices through the Harpers Ferry Historical Association. Films from private organizations and companies are also listed. Contact any agency or organization for a complete catalog of educational films.

Films and Videocassettes

America's Backyard
Type: VHS
Age: High school to adult
Length: 60 minutes
Cost: Free loan
Date: 1991
Source: USDA Forest Service
            Available through Audience Planners, Inc.
            5341 Derry Avenue, Suite PNQ
            Agoura Hills, CA 91301
            (818) 865-1233

This video begins with a brief history of the USFS, with segments on Gifford Pinchot and Smokey Bear, and then continues with separate segments on different aspects of the national forests, including bear research, different types of birds, a pine cone collecting business, llamas and mules as pack animals, and wildlife "hospitals" for wounded animals.

America's Wetlands

Type: 16mm
Age: High school to adult
Length: 28 minutes
Cost: Free loan
Date: 1981
Source: U.S. Fish and Wildlife Service
           Audio Visual Office
           1849 C Street NW
           Washington, DC 20240
           (202) 208-5611

Using scenic photography, music, and narration, this film provides a view of America's wetlands, showing their natural beauty and their importance to birds and other wildlife. The film also discusses the threats to wetlands and what can be done to save them.

Ancient Forests

Type: VHS, 16mm
Age: Grade 7 to adult
Length: 25 minutes
Cost: VHS, $110; 16mm $390
Date: 1992
Source: National Geographic Society
           Educational Services
           P.O. Box 98019
           Washington, DC 20090-8019
           (800) 368-2728

From the redwoods of northern California to Alaska's Tongass National Forest and Prince of Wales Island, this film shows the beauty of these giant thousand-year-old trees and the complex ecosystem of the old-growth forests. The conflicts over management of these resources, and how much should be cut, are presented and discussed.

Ancient Forests: Vanishing Legacy of the Pacific Northwest

Type: VHS
Age: High school to adult
Length: 13 minutes

Cost: $19.95
Date: 1988
Source: The Wilderness Society
900 17th Street, NW
Washington DC 20006-2596
(202) 833-2300

The old-growth forests of the Pacific Northwest are shown in all their splendor, with gigantic, moss covered trees, and ferns across a damp forest floor. In contrast, footage of clear-cut logging operations show the absolute destruction of old-growth areas. Also covered are the local economy, the export of raw logs, and the cancer fighting drug Taxol, produced by the Pacific Yew tree. The film makes a good case for saving the remaining ancient forests in the area.

Arctic National Wildlife Refuge: A Wilderness in Peril
Type: VHS
Age: Grade 7 to adult
Length: 15 minutes
Cost: $19.95
Date: 1993
Source: The Wilderness Society
900 Seventeenth Street NW
Washington, DC 20006-2596
(202) 833-2300

Potential oil and gas development threatens the Arctic National Wildlife Refuge. This film shows the wildlife that thrives there, including the 200,000-head Porcupine Caribou Herd, and the dangers oil development could cause. The lifestyle of the native G'Witchin Indians is also portrayed.

Are We Killing America's Forests?
Type: VHS
Age: High school to adult
Length: 60 minutes
Cost: Free loan
Date: 1989
Source: USDA Forest Service
Available through Audience Planners, Inc.
5341 Derry Avenue, Suite PNQ
Agoura Hills, CA 91301
(818) 865-1233

This video examines threats to forests in America today due to human action and looks at a forest life-cycle. Major forests around the country are used as specific examples.

Balancing Act
Type: VHS
Age: High school to adult
Length: 60 minutes
Cost: Free loan
Date: 1991
Source: USDA Forest Service
            Available through Audience Planners, Inc.
            5341 Derry Avenue, Suite PNQ
            Agoura Hills, CA 91301
            (818) 865-1233

The controversial aspects of USFS policy are discussed by people involved on both sides of the issues, including clear-cut logging, old-growth forests and the northern spotted owl, timber harvesting and sale procedures, livestock grazing, mining operations, and recreational concerns. This video is highly recommended.

Bighorn: Capture and Relocation on California's National Forests
Type: VHS
Age: Grade 9 to adult
Length: 9 minutes
Cost: Free loan
Date: 1990
Source: USDA Forest Service
            Available through Audience Planners, Inc.
            5341 Derry Avenue, Suite PNQ
            Agoura Hills, CA 91301
            (818) 865-1233

Bighorn sheep are captured in one national forest and relocated into another in this brief but engaging film. USFS managers discuss the relocation effort, which is conducted in cooperation with other federal agencies.

By the Grace of Man
Type: VHS, 3/4" video
Age: Junior high school to adult
Length: 20 minutes
Cost: VHS, $335; 3/4" video, $365
Date: 1988
Source: Barr Films
            12801 Schabarum Avenue
            P.O. Box 7878
            Irwindale, CA 91706
            (800) 234-7879

Colorado's Non-Game Wildlife Program is examined in order to demonstrate how conscientious people can help save endangered wildlife, and the impact that wildlife has on other species.

Challenge of Yellowstone

Type:	VHS, 16mm
Age:	High school to adult
Length:	25 minutes
Cost:	VHS, $19.95; 16mm, $342
Date:	1979
Source:	Harpers Ferry Historical Association
	P.O. Box 197
	Harpers Ferry, WV 25425
	(800) 821-5206

Yellowstone was the first park in the U.S. National Park System. This film explains how the idea for Yellowstone originated and provides an introduction to the whole park system as it exists today, including a look at wilderness areas within park boundaries using both early film footage and modern photography.

Denali Wilderness

Type:	VHS, 16mm
Age:	Elementary school to adult
Length:	30 minutes
Cost:	VHS, $19.95; 16mm, $340
Source:	Harpers Ferry Historical Association
	P.O. Box 197
	Harpers Ferry, WV 25425
	(800) 821-5206

Denali National Park is portrayed in footage filmed over the course of a year throughout all four seasons, with scenes of Alaskan bull moose, Dall sheep, caribou, fox, grizzly bear, and wolf. The spectacular scenery of the park and Mount McKinley is also featured.

Do Your Part: A Wetlands Discovery Adventure

Type:	VHS
Age:	Grades 4–6
Length:	20 minutes
Cost:	Free loan
Date:	1992
Source:	USDA Forest Service
	Available through Audience Planners, Inc.
	5241 Derry Avenue, Suite PNQ
	Agoura Hills, CA 91301
	(818) 865-1233

This upbeat video follows young students on an adventure through a wetland and explains why these important ecosystems are disappearing and what can be done to save them.

The Everglades: Conserving a Balanced Community

Type: VHS
Age: Elementary school
Length: 12 minutes
Cost: $99
Date: 1988
Source: Encyclopaedia Britannica Educational Corporation
310 South Michigan Avenue
Chicago, IL 60604
(800) 621-3900

An introduction to plants, animals, and the diverse ecology of the Everglades, with a call for the preservation and conservation of the habitat.

Everglades: Seeking a Balance

Type: 16mm
Age: High school to adult
Length: 16 minutes
Cost: $214
Source: Harpers Ferry Historical Association
P.O. Box 197
Harpers Ferry, WV 25425
(800) 821-5206

The Everglades is an ecosystem in distress, and wildlife studies play a big role in the management of Everglades National Park. This film follows some of those studies and explains the problems faced by those trying to protect this unique area.

Fire and Wildlife: The Habitat Connection

Type: VHS
Age: High school to adult
Length: 50 minutes
Cost: $49.95
Date: 1989
Source: Sponsored by the USDA Forest Service.
Produced by and available through
Stoney-Wolf Productions
304 Main Street
Stevensville, MT 59870
(406) 273-0060

This film first looks at forest fires as a part of the natural forest eco-system, explaining how they affect both the wilderness and wildlife, and then provides a history of the USFS's fire management policies. The original policy of putting out all fires led to high levels of fuel, but many blame their subsequent "let burn" policy for the massive fires of 1988 (see Chapter 1, p. 19–22). Here the narrator explains the conflicts, justifying the "let burn" policy, and showing the previous policies as being more to blame for the size of the 1988 fires. Spectacular footage of those fires is included, along with other wildlife scenes.

For Earth's Sake: The Life and Times of David Brower
Type: VHS
Age: High school to adult
Length: 58 minutes
Cost: $275
Date: 1989
Source: KCTS Television
           Available through Bullfrog Films
           Oley, PA 19547
           (215) 779-8226
           (800) 543-3764

Just as the life of David Brower followed the environmental movement through a period of terrific growth, this film chronicles that movement, and the man responsible for much of that growth. Brower is legendary in environmental circles, and this film provides a portrayal of both the man and his accomplishments, showing in part what one man can do for a cause. An autobiography by the same title was published in 1990 by Peregrine Smith Books.

Forestry: An Overview
Type: VHS, 3/4" video
Age: High school to adult
Length: 20 minutes
Cost: VHS, $149.95; 3/4", $179.95
Date: 1987
Source: Barr Films
           12801 Schabarum Avenue
           P.O. Box 7878
           Irwindale, CA 91706
           (800) 234-7879

The controversies and methods of logging are examined, includ-ing old-growth forests, clear-cutting, conservation, and tree-planting programs. ·

Forests—North American Heritage

Type: VHS
Age: High school
Length: 12 minutes
Cost: $79.95
Date: 1981
Source: West Wind Productions
    P.O. Box 3532
    Boulder, CO 80303
    (800) 228-8854

This film defines a "forest" and looks at the different types of forest in North America, including pine forests of the Pacific Northwest and hardwood forests of the East. Forest management and the future of our forests are also addressed.

Forests in the Balance—A Fight against Time

Type: 16 mm
Age: High school to adult
Length: 27 minutes
Cost: Free loan
Date: 1982
Source: USDA Forest Service
    Available through Audience Planners, Inc.
    5341 Derry Avenue, Suite PNQ
    Agoura Hills, CA 91301
    (818) 865-1233

This film includes a wide variety of forest management topics, including thinning, harvesting, and soil erosion. Much of the film concentrates on integrated pest management (IPM) to control insects and tree diseases. Efforts to eradicate the mountain pine beetle in Montana and the spruce bud worm in Maine are offered as examples.

Forests USA: A Heritage for Our Children

Type: VHS
Age: High school to adult
Length: 10 minutes
Cost: Free loan
Date: 1985
Source: USDA Forest Service
    Available through Audience Planners, Inc.
    5341 Derry Avenue, Suite PNQ
    Agoura Hills, CA 91301
    (818) 865-1233

Beginning with the original colonies, as trees were harvested for wood and burned to clear land, this film follows the history of forests in the

United States. Clear cutting and other present day logging practices are also shown, and conservation management is discussed. The film calls for more controlled management, not just in the United States, but on a worldwide basis.

Fountain of Life

Type:	VHS, 16mm
Age:	Junior high school to adult
Length:	22 minutes
Cost:	VHS, $19.95; 16mm, $284
Source:	Harpers Ferry Historical Association
	P.O. Box 197
	Harpers Ferry, WV 25425
	(800) 821-5206

Rocky Mountain National Park is the subject of this film, with a look at the diversity of plant and animal life found there, and the stunning scenery within the park. A history of farmers, miners, and other early inhabitants in the area is also included.

Giant Sequoia

Type:	VHS, 16mm
Age:	Grade 6 to adult
Length:	17 minutes
Cost:	VHS, $79.95; 16mm, $245
Source:	VHS is available through
	Media Design Associates, Inc.
	Box 3189
	Boulder, CO 80307-3189
	(800) 228-8854
	16mm is available through
	Harpers Ferry Historical Association
	P.O. Box 197
	Harpers Ferry, WV 25425
	(800) 821-5206

In the past decade much has been learned about the biology of the Giant Sequoia. The findings of this research are shown in this film, with a focus on the role fire plays in the Sequoia forest ecosystem.

Giant Sequoia: Past, Present, and Future

Type:	VHS
Age:	Grade 6 to adult
Length:	4 minutes
Cost:	Free loan
Date:	1992

Source: USDA Forest Service
Available through Audience Planners, Inc.
5341 Derry Avenue, Suite PNQ
Agoura Hills, CA 91301
(818) 865-1233

This short video explains the USFS's role in preserving the Giant Sequoia in California, with an overview of historic logging and firefighting practices, an explanation of current protection policy, and a look at plans for restoring young Sequoia.

Glacier Bay: Grand Design
Type: VHS, 16mm
Age: Elementary school to adult
Length: 17 minutes
Cost: VHS, $17.95; 16mm, $239
Source: Harpers Ferry Historical Association
P.O. Box 197
Harpers Ferry, WV 25425
(800) 821-5206

Glacier Bay is known for its dramatic glaciers, spilling out from steep valleys and occasionally falling thunderously into the sea. What is not so readily known is that the retreating glaciers also support an abundance of plant and animal life, including everything from lichens to mature forests. This film portrays the beauty of Glacier Bay National Park and the varied life and terrain found there.

Green Gold
Type: VHS, 16mm
Age: Grade 5 to adult
Length: 12 minutes
Cost: Free loan
Date: 1984
Source: USDA Forest Service
Available through Audience Planners, Inc.
5341 Derry Avenue, Suite PNQ
Agoura Hills, CA 91301
(818) 865-1233

National USFS timber management policies are often criticized by environmentalists. This film presents the USFS's side of the story, with an upbeat look at logging and management policies in the national forests.

Guardians of the Forest
Type: VHS
Age: Grade 7 to adult

Length: 58 minutes
Cost: Free loan
Date: 1984
Source: USDA Forest Service
Available through Audience Planners, Inc.
5341 Derry Avenue, Suite PNQ
Agoura Hills, CA 91301
(818) 865-1233

Narrated by James Earl Jones, this made-for-television special follows USFS managers as they try to locate and eradicate marijuana plantations in the national forests.

Gulf Islands: Beaches, Bays, Sounds, and Bayous
Type: VHS, 16mm
Age: High school to adult
Length: 28 minutes
Cost: VHS, $19.95; 16mm, $381
Source: Harpers Ferry Historical Association
P.O. Box 197
Harpers Ferry, WV 25425
(800) 821-5206

Gulf Islands National Seashore is the subject of this beautifully made film. Specific topics include the formation and physical features of these barrier islands, as well as regional history and recreational opportunities.

In Celebration of America's Wildlife
Type: VHS, Beta, 16mm, 3/4" video, 8mm video
Age: High school and above
Length: 57 minutes (or a 28-minute edited version)
Cost: Beta, VHS, $12; 16mm, $210; 3/4", $20; 8mm video, $20
VHS (28 minutes), $6.50
Date: 1987
Source: Commonwealth Films
1500 Brook Road
Richmond, VA 23220
(804) 649-8611

In 1937 the Pittman-Robertson Act was passed by Congress, creating an excise tax on firearms and hunting equipment to generate funds for wildlife management. This film celebrates the 50th anniversary of the Act by documenting the accomplishments and future prospects of wildlife management in the United States, with scenic wildlife photography and an interesting narration.

In Celebration of Trees

Type: VHS
Age: Grade 6 to adult
Length: 50 minutes
Cost: Free loan
Date: 1991
Source: USDA Forest Service
             Available through Audience Planners, Inc.
             5341 Derry Avenue, Suite PNQ
             Agoura Hills, CA 91301
             (818) 865-1233

Both trees and wildlife are captured with spectacular cinematography in national forests across the country, including forests on the Olympic Peninsula, the Florida Everglades, the Sierra Nevada, and the Shenandoah Valley.

The Incredible Wilderness

Type: 16mm
Age: Elementary school to adult
Length: 10 minutes
Cost: $143
Source: Harpers Ferry Historical Association
             P.O. Box 197
             Harpers Ferry, WV 25425
             (800) 821-5206

Olympic National Park is the subject of this film, with beautiful scenes of the Olympic rain forest, rugged Pacific coastline, wildflower covered alpine meadows in summer, and snowcapped winter peaks.

Into the Wild

Type: VHS
Age: Elementary school
Length: 24 minutes
Cost: $19.95
Date: 1991
Source: National Wildlife Federation
             1400 Sixteenth Street NW
             Washington, DC 20036-2266
             (800) 432-6564

This film is hosted by and produced for kids and provides a look at endangered species, including the red wolf, whooping crane, and humpback whale.

Katmai

Type:	VHS, 16mm
Age:	Elementary school to adult
Length:	16 minutes
Cost:	VHS, $22.95; 16mm, $206
Source:	Harpers Ferry Historical Association
	P.O. Box 197
	Harpers Ferry, WV 25425
	(800) 821-5205

Katmai National Park is one of the most rugged and remote in the National Park System. This film takes the viewer on a tour of this spectacular wilderness, with a look at the scenery, ecology, and wildlife of the area.

Last Stronghold of the Eagles

Type:	VHS
Age:	High school to adult
Length:	30 minutes
Cost:	$250; three-day rental $75
Date:	1981
Source:	Coronet/MTI Film and Video
	(800) 621-2131

The habitat of the declining bald eagle population is shown in Alaska, along with the threats to their existence, and with a look at what may lie ahead for the species in the future.

Legacy for Wings

Type:	VHS, 16mm (closed-captioned version available)
Age:	High school to adult
Length:	30 minutes
Cost:	Free loan
Source:	USDA Forest Service
	Available through Audience Planners, Inc.
	5341 Derry Avenue, Suite PNQ
	Agoura Hills, CA 91301
	(818) 865-1233

As wetlands disappear, the nation's population of duck, geese, and other migratory waterfowl is dropping. This film shows the problems these birds face, and what is being done by the USFS and others to restore wetlands.

The Living Forest

Type:	VHS
Age:	Elementary school

Length: 16 minutes
Cost: $225
Date: 1991
Source: Pyramid Film and Video
            2801 Colorado Avenue
            Santa Monica, CA 90404
            (310) 828-7577

Originally titled *A Walk in the Forest,* this updated version of the film presents the ecology as well as the serenity of the forest and makes a case for conservation and preservation. It shows forest plants and animals during both day and night, and during all four seasons.

Living Waters of the Big Cypress
Type: 16mm
Age: Junior high school to adult
Length: 14 minutes
Cost: $164
Source: Harpers Valley Historical Association
            P.O. Box 197
            Harpers Valley, WV 25425
            (800) 821-5206

The interdependence between plants and animals in the swamp of Big Cypress National Preserve is shown up close with micro-photography and audio of creatures rarely seen by human eyes.

Magnificence in Trust
Type: 16mm
Age: Elementary school to adult
Length: 29 minutes
Cost: $366
Source: Harpers Ferry Historical Association
            P.O. Box 197
            Harpers Ferry, WV 25425
            (800) 821-5206

The Alaskan wilderness is the subject of this beautiful film, featuring scenes of crumbling glaciers falling into Glacier Bay; salmon, brown bears, and smoldering volcanoes at Katmai National Park; and caribou, mountain sheep, moose, and grizzly bear at Mount McKinley in Denali National Park. The film was by Norman G. Dyhrenfurth, world-renowned mountain climber and leader of the first American expedition up Mount Everest.

Managing the Old Growth Forests
Type: VHS; 3/4″ video
Age: High school to adult

Length: 21 minutes
Cost: VHS, $285; ³/₄″, $310
Date: 1987
Source: Barr Films
12801 Schabarum Avenue
P.O. Box 7878
Irwindale, CA 91706
(800) 234-7879

The debate over logging in the old-growth forests of the Pacific Northwest is explained. Both sides agree that some of the old-growth areas must be preserved, but how much? Scientists and logging representatives present their cases.

Mount McKinley—The Land Eternal
Type: 16mm
Age: High school to adult
Length: 24 minutes
Cost: $364
Source: Harpers Ferry Historical Association
P.O. Box 197
Harpers Ferry, WV 25425
(800) 821-5206

The highest mountain in North America, Mount McKinley, is featured in this film, as well as the rest of Denali National Park, in which it is located. All types of life that survive in this harsh climate are shown throughout the year.

National Audubon Society Slide Shows
Type: Slide shows
Age: General and school-age audiences
Length: 14 minutes average
Cost: $50 each
Source: The National Audubon Society
Conservation Information
700 Broadway
New York, NY 10003

The National Audubon Society has produced four separate slide shows to introduce their High Priority Campaigns and educate the public on critical conservation issues. The topics include Arctic National Wildlife Refuge, Endangered Species, Everglades, and Wetlands. They include a cassette tape and script.

National Audubon Society Videos
Type: VHS
Age: High school to adult

Length: 30 to 60 minutes each
Cost: $14.98 (30 minutes); $29.98 (60 minutes), plus shipping and handling
Source: The National Audubon Society
Available through Live Home Video
(800) 782-8226

The National Audubon Society produces excellent television specials which air on public television. These specials are available on VHS, and include the following titles:

Ancient Forests: Rage over Trees

Arctic Refuge: A Vanishing Wilderness?

California Condor

Common Ground: Farming and Wildlife

Ducks Under Siege

Greed and Wildlife: Poaching in America

Grizzly and Man: Uneasy Truce

Messages from the Birds

The Mysterious Black-footed Ferret

Woodstork, Barometer of the Everglades

Wildfire

Most of these films come with a teacher's guide and printed material on the subject and information on how best to use the videos in the classroom.

National Parks: An American Legacy
Type: VHS
Age: High school to adult
Length: 28 minutes
Cost: $19.95
Source: Harpers Ferry Historical Association
P.O. Box 197
Harpers Ferry, WV 25425
(800) 821-5206

The history of the early preservation movement, the establishment of the first national parks, and the NPS is presented here, with information on people who had a significant impact on preservation efforts. The contributions of Stephen Mather, Horace Albright, Franklin and Theodore Roosevelt, and John Muir are remembered, as well as those of more recent NPS officials. Historic photography of the national parks is combined with scenic modern-day footage.

National Parks: Our Treasured Lands
Type: VHS, 16mm
Age: Elementary school to adult
Length: 28 minutes
Cost: VHS, $19.95; 16mm, $392
Source: Harpers Ferry Historical Association
             P.O. Box 197
             Harpers Ferry, WV 25425
             (800) 821-5206

Produced by the NPS, this film actually covers the whole spectrum of units in the park system, from national parks, to national seashores, monuments, rivers, battlefields, and historic sites. It is an excellent introduction to the breadth of the National Park System.

National Parks: Playground or Paradise?
Type: VHS, 16mm
Age: Grade 7 to adult
Length: 59 minutes
Cost: VHS, $90; 16mm, $400
Date: 1981
Source: National Geographic Society
             Educational Services
             P.O. Box 98019
             Washington, DC 20090-8019
             (800) 368-2728

The national parks face a dilemma. These lands are set aside so that the public can enjoy a slice of undisturbed nature, but are we loving our parks to death? The number of annual visitors is seriously damaging the natural ecosystems in the parks, and detracting from the visitor's experience. The film visits Yellowstone, Yosemite, and the Grand Canyon to address this problem. It was originally produced by National Geographic as a television special.

Pa-Hay-Okee: Grassy Waters
Type: VHS, 16mm
Age: Elementary school to adult
Length: 17 minutes
Cost: VHS, $19.95; 16mm, $250
Source: Harpers Ferry Historical Association
             P.O. Box 197
             Harpers Ferry, WV 25425
             (800) 821-5206

Everglades National Park is the subject of this film, with footage of this spectacular grassland and the creatures that live there. The relationship between man and nature is examined with special attention to the ecological problems faced by the Everglades, and attempts to solve them.

Poison in the Rockies

Type: VHS
Age: High school to adult
Length: 54 minutes
Cost: $250
Date: 1990
Source: Earth Image Films
Available through Bullfrog Films
Oley, PA 19547
(215) 779-8226
(800) 543-3764

Water pollution, overpopulation, and acid rain in the Rocky Mountains are the three environmental topics covered by this informative and interesting documentary. A gold rush in the mid and late 1800s brought these problems with it. Surprising to many who view the Rockies as a pristine wilderness, water pollution is among the largest problems facing the Rocky Mountain region. In some areas the local water is not even safe to drink, due to runoff from mining operations that contaminate the surface and ground water with lead, arsenic, cadmium, and other chemicals and minerals. This film includes explanations of the problems and interviews with experts and government officials, backed up with impressive photography.

A Precious Legacy: Our National Forest System

Type: VHS
Age: Grade 7 to adult
Length: 5 minutes
Cost: Free loan
Date: 1991
Source: USDA Forest Service
Available through Audience Planners, Inc.
5241 Derry Avenue, Suite PNQ
Agoura Hills, CA 91301
(818) 865-1233

This video is a brief yet informative introduction to the national forests and the history and mission of USFS, and uses both historic photographs and modern film footage.

Problems of Conservation: Acid Rain

Type: VHS
Age: Junior high school to adult

Length: 18 minutes
Cost: $99
Date: 1990
Source: Encyclopaedia Britannica Educational Corporation
310 South Michigan Avenue
Chicago, IL 60604
(800) 554-9862

The causes of acid rain are examined, with a look at the specific problems it creates, from damaged trees and soils, to diminishing fish populations. The video also explains steps that can be taken to reduce the threat of acid rain.

Protecting Endangered Animals
Type: VHS, 16mm
Age: Grades 4 to 9
Length: 15 minutes
Cost: VHS, $56; 16mm, $185
Date: 1984
Source: National Geographic Society
Educational Services
P.O. Box 98019
Washington, DC 20090-8019
(800) 368-2728

Animals that are already extinct are discussed, from dinosaurs to passenger pigeons, along with the reasons for their demise. Endangered animals from the present are also shown, including bald eagles, black-footed ferrets, and manatees. The effects of humans on these species are documented, with a look at how mankind is causing their demise, and what some people are doing to try to save them.

Sacred Trust: Preserving America's Wilderness
Type: VHS
Age: High school to adult
Length: 18 minutes
Cost: $19.95
Date: 1988
Source: The Wilderness Society
900 17th Street, NW
Washington DC 20006-2596
(202) 833-2300

Narrated by Robert Redford, this film presents the history of the wilderness preservation movement, as well as the history of The Wilderness Society. It covers the signing of the Wilderness Act of 1964, clear-cut logging, mining, and the national parks, with beautiful scenic footage.

A Sauk County Almanac

Type: VHS, ³/₄″ video
Age: High school to adult
Length: 28 minutes
Cost: VHS, $355; ³/₄″, $385
Date: 1989
Source: Barr Films
12801 Schabarum Avenue
P.O. Box 7878
Irwindale, CA 92706
(800) 234-7879

In 1952, Aldo Leopold wrote about Sauk County in his famous book, *A Sand County Almanac*. This film returns to the area to show the bald eagles, turkey vultures, geologic formations, and native prairies found there, and calls on viewers to redefine their ideas about the natural world and their place in it.

Save the Wetlands

Type: VHS, ³/₄″ video
Age: High school to adult
Length: 20 minutes
Cost: VHS, $149.95; ³/₄″, $179.95
Date: 1986
Source: Barr Films
12801 Schabarum Avenue
P.O. Box 7878
Irwindale, CA 91706
(800) 234-7879

The ecology of wetlands is examined, with a look at the threats to these systems, including farm runoff, draining, and development. The film increases an awareness of the need to protect wetlands and rivers.

Schedadxw (Cha-Da'-Duch)

Type: VHS, 16mm (closed-captioned version available)
Age: High school to adult
Length: 30 minutes
Cost: Free loan
Date: 1987
Source: USDA Forest Service
Available through Audience Planners, Inc.
5341 Derry Avenue, Suite PNQ
Agoura Hills, CA 91301
(818) 865-1233

Salmon and steel-head trout populations in the Pacific Northwest have fallen dramatically during the past several decades due to dam building and silt runoff from logging operations. This film shows the problems these fish face and presents possible solutions.

Shenandoah: The Gift

Type: VHS, 16mm
Age: Junior high school to adult
Length: 17 minutes
Cost: VHS, $19.95; 16mm, $251
Source: Harpers Ferry Historical Association
P.O. Box 197
Harpers Ferry, WV 25425
(800) 821-5206

Before the Shenandoah Valley was made into a national park, the area was populated, farmed, and heavily logged. This film tells the story of the valley's purchase by the federal government and the transformation back to its original, natural state, using both historic and modern photographs and film footage.

Showcase of the Ages

Type: VHS
Age: Elementary school
Length: 41 minutes
Cost: $195
Date: 1988
Source: Pyramid Film and Video
2801 Colorado Avenue
Santa Monica, CA 90404
(310) 828-7577

The geology of the Grand Canyon is the subject of this in-depth documentary, featuring scenic cinematography. The viewer follows along on a two-week raft trip down the Colorado River along with a noted geologist.

A Swamp Ecosystem

Type: VHS, 16mm
Age: Grade 7 to adult
Length: 23 minutes
Cost: VHS, $79; 16mm $101.50
Date: 1983

Source: National Geographic Society
             Educational Services
             P.O. Box 98019
             Washington, DC 20090-8019
             (800) 368-2728

The ecosystem of the Okefenokee Swamp is explored, with special attention to the interaction among plant and animal life, from alligators to carnivorous pitcher plants. Special attention is paid to the positive contributions that drought and fire make to the swamp's long-term health.

T. R. Country
Type: VHS, 16mm
Age: Elementary school to adult
Length: 15 minutes
Cost: VHS, $16.95; 16mm, $187
Source: Harpers Ferry Historical Association
             P.O. Box 197
             Harpers Ferry, WV 25425
             (800) 821-5206

Theodore Roosevelt lived for a time in the Badlands of North Dakota. He came to love this rugged territory, and his time here shaped his conservation philosophy to a large extent. The deserts, valleys, prairies, and plateaus are displayed in vivid film footage during both the winter and summer seasons, with quotes from Roosevelt.

This Island Earth
Type: VHS
Age: Elementary through high school
Length: 48 minutes
Cost: $14.95, and $1.95 (shipping and handling)
Date: 1992
Source: The National Audubon Society
             (800) 989-0227

This is part music video, with narrated sequences by Kenny Loggins about how humans endanger the earth and its species, and offering tips to help the earth's ecology. It features the music of Loggins, Gloria Estefan, and Shanice.

This Land Is Your Land
Type: VHS
Age: Grade 6 to adult
Length: 60 minutes
Cost: Free loan
Date: 1990

Source: USDA Forest Service
Available through Audience Planners, Inc.
5341 Derry Avenue, Suite PNQ
Agoura Hills, CA 91301
(818) 865-1233

An overview of the National Forest System, this video provides information on the forests themselves, as well as budgets, employees, and the organizational structure of the USFS. Discussions are included on the more controversial aspects of managing the national forests, including wilderness, wildlife, economic, and recreational concerns.

A Treasure in the Sea
Type: VHS, 16mm
Age: Elementary to high school
Length: 24 minutes
Cost: VHS, $19.95; 16mm, $336
Source: Harpers Ferry Historical Association
P.O. Box 197
Harpers Ferry, WV 25425
(800) 821-5206

The history and natural history of islands within Channel Islands National Park are the subject of this film. These islands contain the last coastal wilderness in southern California, with abundant sea life, wildlife, birds, and plants. The footage includes a look at bleak caliche forests, in which natural sand casts have been made of once living plants by the forces of time, wind, the sun, and chemical change.

Up from the Ashes
Type: VHS
Age: Grade 7 to adult
Length: 15 minutes
Cost: Free loan
Date: 1989
Source: USDA Forest Service
Available through Audience Planners, Inc.
5341 Derry Avenue, Suite PNQ
Agoura Hills, CA 91301
(818) 865-1233

After forest fires the USFS has the task of rehabilitation in order to prevent soil erosion, protect watersheds, salvage timber, and reforest the area. The activities of USFS employees and volunteers, including school children, are documented in this video about the aftermath of fires in 1987.

Up in Flames: A History of Firefighting in the Forest

Type: VHS
Age: High school to adult
Length: 29 minutes
Cost: Free loan
Date: 1981
Source: USDA Forest Service
Available through Audience Planners, Inc.
5341 Derry Avenue, Suite PNQ
Agoura Hills, CA 91301
(818) 865-1233

An overview of the history and evolution of firefighting in the forest is presented, from the first historic fires to the modern day. Firefighting tools, communications systems, and organizations are shown with historical and modern footage. This video was produced by the Forest History Society.

Visions of the Wild

Type: VHS, 16mm (closed-captioned version available)
Age: Grade 7 to adult
Length: 23 minutes
Cost: Free loan
Date: 1985
Source: USDA Forest Service
Available through Audience Planners, Inc.
5341 Derry Avenue, Suite PNQ
Agoura Hills, CA 91301
(818) 865-1233

This history of wilderness and the wilderness preservation movement focuses on the national forests and USFS management policies.

Wetlands Ecology: Estuaries

Type: VHS, ³/₄″ video
Age: High school and college
Length: 23 minutes
Cost: VHS, $295; ³/₄″, $325
Date: 1991
Source: Barr Films
12801 Schabarum Avenue
P.O. Box 7878
Irwindale, CA 92706
(800) 234-7879

An estuary is a wetland between an ocean and fresh waters. This film uses estuaries in Washington state to demonstrate the ecology and diversity of life in this type of ecosystem.

What Remains for the Future?

Type: VHS, 3/4" video
Age: High school and college
Length: 28 minutes
Cost: VHS, $295; 3/4", $325
Date: 1991
Source: Barr Films
12801 Schabarum Avenue
P.O. Box 7878
Irwindale, CA 92706
(800) 234-7879

The history and ecology of the Pacific Northwest is the subject of this video, including the plants and animals that live in the area, as well as the differences between mountain and coastal environments, and the impact of recent human logging and development.

Wild by Law

Type: VHS
Age: High school to adult
Length: 52 minutes
Cost: Free loan
Date: 1991
Source: USDA Forest Service
Available through Audience Planners, Inc.
5341 Derry Avenue, Suite PNQ
Agoura Hills, CA 91301
(818) 865-1233

This video follows the story behind the passage of the Wilderness Act of 1964, with a look at the three men most responsible for passage of the act: Aldo Leopold, Bob Marshall, and Howard Zahniser. It was originally produced as part of a series on PBS called *The Wilderness Idea.*

Wild Waters . . . Harmony, Balance, and Use

Type: VHS, 16mm
Age: High school to adult
Length: 15 minutes
Cost: Free loan
Date: 1984

Source: USDA Forest Service
Available through Audience Planners, Inc.
5341 Derry Avenue, Suite PNQ
Agoura Hills, CA 91301
(818) 865-1233

Watershed management in the national forests is covered here, including watershed rehabilitation, restoration, and water quality maintenance.

The Wilderness Idea: John Muir, Gifford Pinchot, and the First Great Battle for Wilderness
Type: VHS
Age: High school to adult
Length: 58 minutes
Cost: Free loan
Date: 1990
Source: USDA Forest Service
Available through Audience Planners, Inc.
5341 Derry Avenue, Suite PNQ
Agoura Hills, CA 91301
(818) 865-1233

Also part of *The Wilderness Idea* series on PBS, this video traces the battle over the Hetch Hetchy valley in Yosemite National Park between conservationists and preservationists. Gifford Pinchot, the first head of the USFS represented conservationists, who believed that building a dam to flood the valley and create a reservoir was a wise idea. Muir and the preservationists were horrified at the thought of losing this beautiful valley beneath a man-made lake. Eventually it was Pinchot who was victorious, and the video visits the site of the reservoir today.

Wildlife—An American Heritage
Type: VHS
Age: Grade 6 to adult
Length: 14 minutes
Cost: $79.95
Date: 1977
Source: Media Design Associates
Box 3189
Boulder, CO 80307-3189
(800) 228-8854

The value of America's diverse wildlife is considered, with outstanding wildlife photography. Basic concepts in wildlife management are examined.

Yellowstone, The Unfinished Song

Type: 16mm
Age: Junior high school to adult
Length: 20 minutes
Cost: $273
Source: Harpers Valley Historic Association
P.O. Box 197
Harpers Valley, WV 25425
(800) 821-5206

During the summer of 1988, forest fires raged out of control over many western states. Yellowstone National Park burned out of control for over three months. The fires themselves are chronicled here, as well as the forests' recovery over the ensuing year, up until the spring of 1989, with a look at how a forest ecosystem is able to regenerate.

Databases

The following is a list of databases relating to the subject of wilderness preservation and conservation. The different databases contain a range of information, from bibliographies and statistics, to the complete text of newspaper and journal articles. Most are available online, which means that a personal computer with a modem and the correct software can directly access the information over telephone lines. Many are also available on diskettes, CD-ROM, or magnetic tape. User fees for these services vary considerably. For current information on fees and content of a particular database, contact that company's customer service representative.

Acid Rain
Bowker A & I Publishing
Reed Reference Publishing Group
121 Chanlon Road
New Providence, NJ 07974
(908) 464-6800
(800) 521-8110
FAX (908) 665-3528

This database covers all aspects of acid rain on a global level, including scientific research, health issues, and U.S. public policy. It contains bibliographical citations and abstracts of journal articles, government and business reports, books, and conference proceedings.

Availability: Online $115.79/connect hour. Also available on magnetic tape and in published form as *Acid Rain Abstracts*.

BNA California Environment Daily

The Bureau of National Affairs, Inc. (BNA)
1231 25th Street, NW
Washington, DC 20037
(202) 452-4132
(800) 452-7773
FAX (202) 822-8092

The Bureau of National Affairs produces 20 publications and databases on topics relating to the federal government, from financial services, to public policy, daily news, and several publications on the environment. The California Environment Daily is a newsletter and database following legal and legislative topics concerning the environment in California, including state and federal court decisions and pending cases. The database contains issues dating back to 1991.

Availability: Online through LEXIS (BNACED), HRIN, The Human Resource Information Network (a daily developments file), and Westlaw (BNA-CED). Also available as a newsletter, and as part of the BNA Daily News database.

BNA Daily Environment Report

The Bureau of National Affairs, Inc. (BNA)
1231 25th Street NW
Washington, DC 20037
(202) 452-4132
(800) 452-7773
FAX (202) 822-8092

The BNA Daily Environment Report is produced as a newsletter and a database, and includes full articles and information, updated daily, on all types of environmental issues in the United States and the court decisions and public policy involved.

Availability: Online through the Human Resource Information Network (HRIN). Also available as a daily newsletter.

BNA Environmental Law Update

The Bureau of National Affairs, Inc. (BNA)
1231 25th Street, NW
Washington, DC 20037
(202) 452-4132
(800) 452-7773
FAX (202) 822-8092

All legal actions and court decisions that affect U.S. environmental policy are covered by this database, which is updated daily. Topics include, among others, wetlands delineation, challenges to the Endangered Species Act, Superfund cleanup, acid rain, and the northern spotted owl. A separate BNA database, BNA Environmental Law Database, concentrates on pesticides.
Availability: Online through LEXIS (BNAELU) and WESTLAW (BNA-ELU). Also available as a newsletter.

Down to Earth!—Close-ups of Our Natural Environment
Wayzata Technology, Inc.
P.O. Box 807
Grand Rapids, MN 55744
(218) 326-0597
(800) 735-7321
FAX (218) 326-0598

More than 750 color and black and white pictures make up this database, including images of scenery and close-ups of flowers, sea shells, fruits and vegetables, leaves, trees, and other natural items. The images can be exported into a desktop publishing program and then edited and printed.
Availability: CD-ROM for Apple Macintosh Plus, SE, or II series with Apple CD SC or compatible CD-ROM drive. Hard disk with 1MB memory recommended. Price: $249.

ENFLEX INFO
ERM Computer Services, Inc.
855 Springdale Drive
Exton, PA 19341
(215) 524-3600
(800) 365-2146
FAX (215) 524-4801

This database contains the full text of U.S. state and federal environmental regulations, and is produced primarily for businesses whose activities are regulated by those regulations.
Availability: CD-ROM only. Requires IBM PC or compatible with 640K memory, 1MB hard disk drive, and a Hitachi CD-ROM drive. $2,500 for 1 jurisdiction, $500 for each additional jurisdiction.

Enviroline
Bowker A & I Publishing
Reed Reference Publishing Group
121 Chanlon Road
New Providence, NJ 07974
(908) 464-6800
(800) 521-8110
FAX (908) 665-3528

Bibliographical information and abstracts of printed matter are contained in this database. Sources include newspaper, magazine, and journal articles, research reports, conference papers, government documents, and books. Topics include all aspects of the environment.
Availability: Online for $114 to $120/connect hour through Data-Star (ENVN), DIALOG Information Services, Inc., DIMDI (EL), ORBIT Online, Life Science Network, and VIZ Technik (ENVN). Also available on CD-ROM, magnetic tape, and in published form as Environmental Abstracts.

Environment News Service

Environment News Service
2488 A Waiomao Road
Honolulu, HI 96816

This database service first began operation in 1992 and provides articles on a variety of environmental issues, including land use, endangered species, changing climates, environmental politics, and wilderness conservation.
Availability: Online through NewsNet, Inc. (EV45).

Environment Reporter

The Bureau of National Affairs, Inc. (BNA)
1231 25th Street, NW
Washington, DC 20037
(202) 452-4132
(800) 452-7773
FAX (202) 822-8092

The Environment Reporter is a weekly newsletter and database (with current and back issues) covering all aspects of the environment with a concentration on the legal, judicial, and public policy aspects of the issues. Topics include hazardous waste, pollution, land use, and conservation of natural resources.
Availability: Online through HRIN, The Human Resources Information Network (ER, a daily developments file), LEXIS (ENVREP), and WEST-LAW (BNA-ER). Also available as part of the BNA Environmental Law Database and as a weekly newsletter.

Environment Week

King Communications Group, Inc.
627 National Press Building
Washington, DC 20045
(202) 638-4260
FAX (202) 662-9744

Environment Week is a newsletter and database containing information on environmental issues and following the actions of U.S. government departments and agencies. The database contains complete copies of the newsletter dating back to 1989.

Availability: Online through NewsNet, Inc. (EV25): $108/connect hour for subscribers to the print publication, $156/connect hour for non-subscribers.

Environmental Bibliography

International Academy at Santa Barbara
Environmental Studies Institute
800 Garden Street, Suite D
Santa Barbara, CA 93101-1552
(805) 965-5010
FAX (805) 965-6071

This database contains over 450,000 bibliographical references to books and magazine and journal articles covering all aspects of the environment, including laws and regulations, energy, wildlife habitats, wilderness preservation, forestry, soil erosion and conservation, agriculture, and pollution.

Availability: Online through DIALOG Information Services, Inc. $60/connect hour, 15 cents per record offline. Also available on CD-ROM and in published form as the *Environmental Periodicals Bibliography*.

Environmental Law Reporter (ELR)

Environmental Law Institute
1616 P Street NW, Suite 200
Washington, DC 20036
(202) 328-5150
FAX (202) 328-5002

The *Environmental Law Reporter* is a monthly journal, the full text of which is also available in database format. The journal covers environmental law and public policy, and is divided into the following eight units:

ELR Administrative Materials (ELR-ADMIN) Includes the text of government environmental law documents relating to such topics as federal regulations, enforcement, and EPA Administrative Law Judge decisions.

ELR Bibliography (ELR-BIB) Includes bibliographical references to law articles from legal journals.

ELR Litigation (ELR-LIT) Includes the text of court decisions, both state and federal, that relate to environmental issues.

ELR News and Analysis (ELR-NEWS) Includes news and editorials on cases in federal and state courts, legislation in Congress, and federal agency policy.

ELR Pending Litigation (ELR-PEND) Includes information on federal and state cases that have not yet been decided in court.

ELR Statutes and International Agreements (ELR-STIA) Includes the complete text of international environmental treaties and agreements, and domestic environmental law statutes.

ELR Superfund (ELR-SF) The Comprehensive Environmental Response, Compensations and Liability Act of 1980, known as the Superfund, was developed to clean up toxic dump sites. This section of the database contains references to the Superfund.

ELR Update (ELR-UPDATE) The ELR produces weekly newsletters covering congressional legislation, government policy, and environmental law cases. ELR update contains the complete text of those newsletters.

Availability: Online through WESTLAW and LEXIS: $48/connect hour, 20 cents to 70 cents per line displayed or $6 to $91 per search, depending on the database. Call for specific information.

GAIA Environmental Resource

Wayzata Technology, Inc.
P.O. Box 807
Grand Rapids, MN 55744
(218) 326-0597
(800) 735-7321
FAX (218) 326-0598

This is a database that contains 400 high-resolution, 24-bit color pictures of nature scenes. The database includes information on the environment and environmental organizations and publications. The pictures include landscapes, forests, logging, flowers, fishing, and the sea.
Availability: Apple Macintosh CD-ROM drive: $249. Not available online.

Guide to Federal Environmental Laws

The Bureau of National Affairs, Inc. (BNA)
1231 25th Street, NW
Washington, DC 20037
(202) 452-4132
(800) 452-7773
FAX (202) 822-8092

The Guide to Federal Environmental Laws is a summarization of the major environmental laws. The text was current in 1991, but is not updated, so any laws passed after that time are not included.
Availability: Online through HRIN, The Human Resource Information Network (a Special Reports Library file). Also available in printed form.

National Environmental Data Referral Service (NEDRES) Database
U.S. National Environmental, Satellite, Data, and Information Service (NESDIS)
National Environmental Data Referral Service (NEDRES)
1825 Connecticut Avenue, NW
Washington, DC 20235
(202) 606-4548
FAX (202) 673-0509

This service provides scientific data on a wide range of environmental topics, collected from sources including weather stations, oceanographic vessels and buoys, satellites, and scientific observers.
Availability: Online through BRS Online (NEDS): $46/connect hour with discounts available through an Advance Purchase plan, BRS/After Dark (NEDS): $15/connect hour, BRS/COLLEAGUE (NEDS): $56 connect/ hour prime-time, $46/connect hour non prime-time.

Network Earth Forum
Turner Broadcasting System (TBS)
1 CNN Plaza
Atlanta, GA 30348
(404) 827-1700

This database provides information from a continuing TBS television program on the environment. It includes environmental news, as well as tips for being environmentally conscious at home, by means such as recycling and conserving water and energy.
Availability: Online through CompuServe Information Service.

Outdoor Forum
CompuServe Information Service
5000 Arlington Centre Boulevard
P.O. Box 20212
Columbus, OH 43220
(614) 457-8600
(800) 848-8199
FAX (614) 457-0348

The Outdoor Forum is a bulletin board on which outdoor enthusiasts can communicate with each other and with various environmental

organizations. The database also contains information on parks and campgrounds, outdoor recreation, wildlife, and conservation.
Availability: CompuServe Information Service (OUTDOORFORUM): $12.80/connect hour (1200 and 2400 baud); $22.80/connect hour (9600 baud).

PressNet Environmental Reports

PressNet Systems, Inc.
400 East Pratt Street, 8th Floor
Baltimore, MD 21202
(410) 625-4998

This database contains abstracts of U.S. and Canadian newspaper articles relating to environmental policy and actions taken by state, local, and federal government. The coverage includes articles from 1989 to the present, but it is not updated on a regular basis.
Availability: Online through PressNet Systems, Inc.

Wildlife Review/Fisheries Review

U.S. Fish and Wildlife Service
C Street between 18th and 19th Streets, NW
Washington, DC 20240
(202) 343-4717

The Wildlife Review and *Fisheries Review* are available together on CD-ROM, or separately in published form. Together they contain nearly 300,000 bibliographic entries and abstracts on fish and wildlife, taken from scientific journals, government reports, conference proceedings, and graduate dissertations. *The Wildlife Review* is also available as part of the *Wildlife Worldwide Database* (next entry).
Availability: CD-ROM.

Wildlife Worldwide

National Information Services Corporation (NISC)
Wyman Towers, Suite 6
3100 St. Paul Street
Baltimore, MD 21218
(410) 243-0797
FAX (410) 243-0982

The National Information Services Corporation works in conjunction with the Wildlife Information Service and the USFWS to produce this bibliographic database. Books, articles, speeches and other literature on the world's wildlife is cited with abstracts. The database is divided into three files:

BIODOC Scientific literature citations cataloged by the Biological Documentation Center at the National University of Costa Rica.

H.E.R.M.A.N. Scientific citations and abstracts cataloged by the Wildlife Information Service, with special attention to marine birds and mammals.

Wildlife Review Over 200,000 citations and abstracts to a wide range of literature on all types of wildlife, habitats, wildlife management, hunting, economics, and policy. This file is also available in published form as *The Wildlife Review*.

Availability: CD-ROM on IBM PC or compatibles only. Requires a hard disk with at least 1MB of memory, MS-DOS, monochrome or color monitor, MS-DOS CD-ROM extensions, and a CD-ROM drive. Price: $695 for an annual subscription.

World Environment Outlook
Business Publishers, Inc. (BPI)
951 Pershing Dr.
Silver Spring, MD 20910-4464
(301) 587-6300
FAX (301) 587-1081

The World Environment Report is a bimonthly newsletter with articles on environmental topics worldwide, including land use planning and conservation, pollution, natural resources, and the greenhouse effect. The *World Environment Outlook* is a database containing complete issues of the newsletter dating back to 1981.

Availability: Online through NewsNet, Inc. (EVOL): $60/connect hour for subscribers to the print publication, $252/connect hour for non-subscribers.

Glossary

acid rain   Air pollution, in the form of sulfur dioxide and nitrogen oxide, is transformed into acidic sulfates and nitrates when transported in rain clouds. These components come to the ground as acid precipitation (rain or snow) and can cause serious damage to aquatic environments, killing off fish and other species.

acre   One acre is a measurement of land equivalent to 43,560 square feet, or a square parcel of land 270 feet on each side.

annual ring   Trees add a ring each year, indicating their growth in width for that year. The ring can be seen in the cross section of the trunk once the tree has been cut. The number of rings corresponds to the age of the tree in years.

arid   Dry, parched.

biomass   The dry weight of all living matter within a specific area.

board foot   A measurement of wood equaling 12 x 12 x 1 inches.

boreal   Northern.

canopy   The upper level of leaves and branches that face the sun and form the top of a forest.

classified wilderness   Lands formally protected in the United States by the Alaska National Interest Lands Conservation Act (ANILCA) of 1980, the Federal Land Policy and Management Act of 1976, and/or the Wilderness Act of 1964.

clear cutting   A method of logging in which all standing trees in a given area are cut, after which the remaining undergrowth is often burned.

climax ecosystem   A mature ecosystem, in a state of balance.

climax forest   A mature forest ecosystem.

coniferous tree A tree whose seeds are carried within cones, such as various pine trees. There are between 500 and 600 species of coniferous trees and shrubs.

conservation Planning the use of land and resources in a way that assures a constant supply of resources for the future.

de facto wilderness Undeveloped, roadless public areas which fit a general description of wilderness, but do not as of yet have any specific legislative protection or wilderness classification.

de jure wilderness Areas specifically classified and protected by the Wilderness Act of 1964.

deciduous A tree or plant that loses its leaves during a certain period of the year, usually winter.

desert An arid region with low rainfall, inhabited by plants and animals specially adapted to living with minimal amounts of water.

designated wilderness. A wilderness area, voted upon and established by the U.S. Congress.

duff Organic matter on a forest floor that is in a state of decomposition.

ecology The science pertaining to the relationship between and among plants and animals.

ecosystem An environment in which all organisms, plant and animal species, exist in balance with one another; a specific natural environment.

effluent A discharge of contaminated material. Effluent can take the form of polluted water emptying from an industrial plant, or smoke from a smokestack.

endangered species A plant or animal species in danger of becoming extinct in the near future.

Environmental Impact Statement (EIS) A study and report which may be required before any development or activities can be performed on an environmentally sensitive area.

erosion The wearing away of surfaces due to abrasion from water, wind, or wind-borne substances.

estuary An inland channel open to the ebb and flow of sea tides. Estuaries are commonly part of a river, and are important for many aquatic organisms and waterfowl.

eutrophic A body of water with excessive nitrates, phosphates, or other plant nutrients.

eutrophication The state in which an excess of plant nutrients favors plant life over animal life and lowers the oxygen content in the water. In some areas eutrophication can also favor nonnative plants over native ones.

extinct A species of plant or animal that has completely died out is extinct.

habitat The place where a plant or animal lives. For instance, Everglades National Park, Florida, is a habitat which supports alligators, fish, and various bird species.

hard release Lands which have been considered but not selected for wilderness protection. These lands are released for multiple-use purposes.

indigenous People, plants, and animals native and occurring naturally in a given area.

inholding A tract of land completely surrounded by public lands, but under private ownership.

multiple use A term used most often by the USFS for their policy of using forests for many purposes, including timber harvesting, recreation, and wilderness preservation.

national forest A forest under control of the USFS. National forests are managed for recreation, timber harvesting, and preservation purposes.

national park A park designated as such, owned by the federal government, and managed by the National Park Service for preservation of the area and recreation purposes.

national wildlife refuge A specific tract of land within the National Wildlife Refuge System, and under the jurisdiction and management of the USFWS. National Wildlife Refuges are managed with the well-being of the area's wildlife as the primary concern.

NPS The U.S. National Park Service, an agency within the Department of the Interior.

old-growth forest A fully developed, mature forest, usually at least 200 years old.

perennial A plant that lives and produces seeds every year for several years.

pH A numeric measure of relative alkalinity or acidity, on a scale from 0 to 14. Acids have pH values below 7 on the scale, 7 is neutral, and a value above 7 indicates alkalinity.

photosynthesis The process by which plants use sunlight to convert carbon dioxide into water and sugar.

prairie A tract of level or rolling grassland.

preservation Setting aside an area for complete protection in its natural state.

public domain lands This term is most often used to describe the western lands obtained by the U.S. government through purchase or treaty during the 1800s. Occasionally the term is used synonymously with public lands.

public lands Any lands owned by the U.S. government.

rangeland Open land supporting grass or shrubs. Cattle are frequently grazed on rangeland.

Roadless Area Review and Evaluation (RARE) An inventory of roadless areas on USFS lands, taken in the early 1970s. A second review (RARE II) was completed from 1977–1979, and designated 2,919 roadless areas with a total of 62 million acres. The studies analyzed each area to determine the best possible use, such as mining, grazing, logging, or preservation as a wilderness area.

savannah (or savanna) A tropical or subtropical grassland (as in Florida), or an open, treeless plain.

second generation forest A forest made up of trees that have been planted to replace those cut during clear-cut logging.

silvics The study of the life history of forest trees. By studying the width of a tree's annual rings, for example, scientists can determine which years were wet and which dry.

silviculture Applying the use of silvics to manage the establishment, growth, and composition of a forest.

soft release Lands considered for and not selected as wilderness areas at a given time, but left open for further consideration in the future.

species A group of animals or plants which possess one or more common characteristics which separate them from other species.

threatened species   A species on the verge of becoming endangered in the near future due to declining numbers.

tree line   (or timber line)   The farthest limit, either in altitude on a mountain, or the farthest north in the northern hemisphere, in which trees are able to grow. Beyond this line, the environment is too harsh for trees to survive.

tundra   A treeless plain, found in the polar regions, characterized by a frozen subsoil, or permafrost, covered with a dark soil and a dense growth of mosses and lichens.

virgin forest   A forest which has never been logged or disrupted by humans.

waterfowl production area   A waterfowl breeding area set aside and managed by the U.S. Fish and Wildlife Service.

wetland   An area covered by shallow water for a portion of each year.

wilderness   An area in which humans have had little or no impact on the ecology, and in which no humans reside.

wilderness area   An uninhabited and undeveloped area that the U.S. Congress has voted to grant special status and protection under authority of the Wilderness Act of 1964.

Wilderness Preservation System   The total sum of lands designated wilderness areas by Congress, under the authority of the Wilderness Act of 1964.

Wilderness Study Area (WSA)   An area that has been mandated for study to determine whether or not it classifies as a wilderness area. This term was originally used by the BLM during a wilderness review in 1978–1980.

wildlife   Any plant or animal species which has not been domesticated, but which lives in the wild.

wise use   A term used to describe the conservationist philosophy of making the most practical use of wild areas on a limited basis, such as through limited logging, dam building, or preservation. The problem with this term is that what one person may consider wise, another may consider foolish.

Index

Childhood summers hiking with his father in the Sierra Nevada mountains gave Kenneth Rosenberg a deep appreciation for the wilderness. The rapid disappearance of open space in his native Southern California gave him a concern for preservation. Mr. Rosenberg is a freelance writer based in Laguna Beach, California.